The Belt and Road

Sketch of the Belt and Road and the Cities along the Economic Corridor

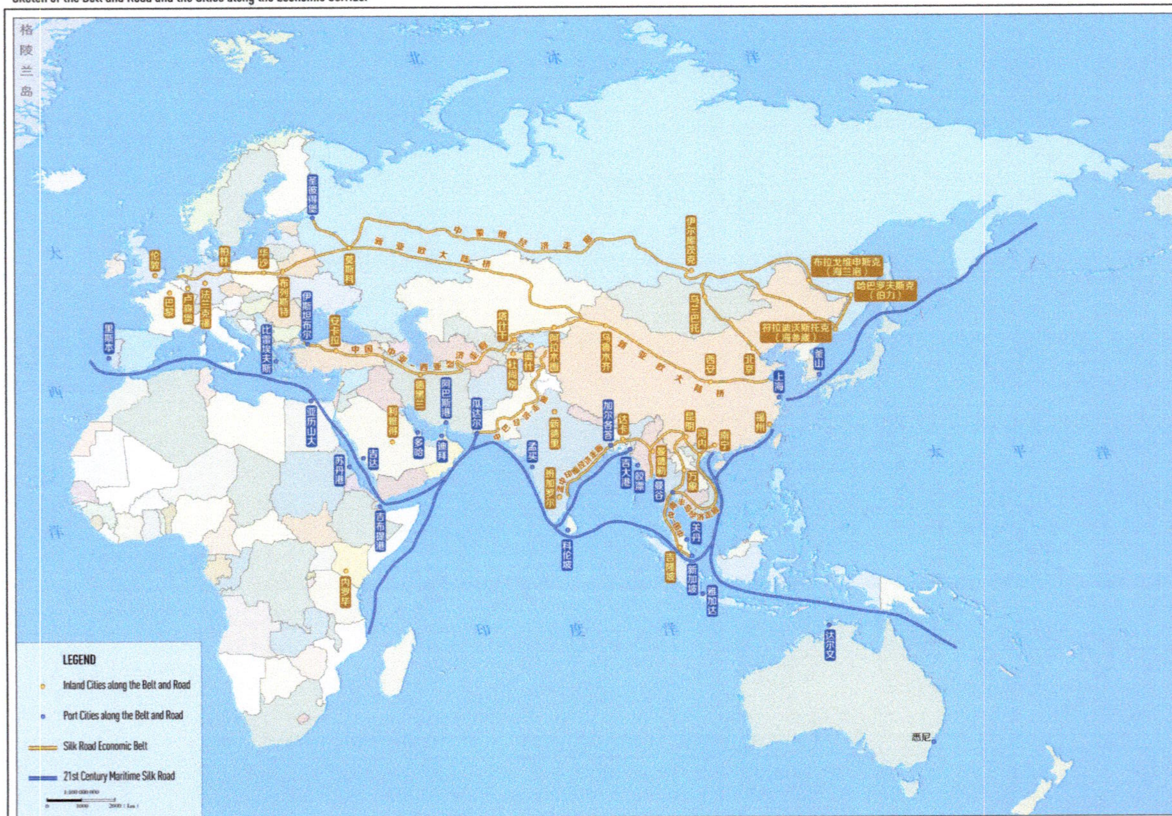

审图号：GS(2016)1766号

国家测绘地理信息局 监制

Fei Xu

The Belt and Road

The Global Strategy of China High-Speed Railway

格致出版社
Truth & Wisdom Press

Springer

Fei Xu
Southwest Jiaotong University
Chengdu, Sichuan, China

ISBN 978-981-13-1104-8 ISBN 978-981-13-1105-5 (eBook)
https://doi.org/10.1007/978-981-13-1105-5

Jointly published with Truth & Wisdom Press, Shanghai, China

The print edition is not for sale in China Mainland. Customers from China Mainland please order the print book from: Truth & Wisdom Press.

Library of Congress Control Number: 2018945457

Printed on acid-free paper

This Springer imprint is published by the registered company Springer Nature Singapore Pte Ltd.
The registered company address is: 152 Beach Road, #21-01/04 Gateway East, Singapore 189721, Singapore

Contents

Part I The Global Strategic Value of China High-Speed Railway

1 China High-Speed Railway: Country's Golden Name Card 3
1.1 Three Climaxes of HSR Construction in the World 3
1.1.1 The First Wave (From the 1960s to the End of 1980s) . . . 3
1.1.2 The Second Wave (From the End of 1980s to the Mid-1990s) 4
1.1.3 The Third Wave (From the Mid-1990s Up to Now) 4
1.2 Overtaking on Curved Lane: Innovation History of CHSR 4
1.2.1 Technology Accumulation Period (Before 2004) 5
1.2.2 Period for Technology Import, Digestion, Absorption and Re-innovation (2004–2008) 6
1.2.3 Indigenous Innovation Period (2008–2012) 6
1.2.4 All-Round Indigenous Innovation Period (Since 2012) . . . 7
1.3 Achievements of CHSR that Attract World Attention 12
1.3.1 The Largest Scale 12
1.3.2 Advanced Technologies 13
1.3.3 Safe and Reliable 14
1.3.4 Good Compatibility 14
1.3.5 High Performance to Price Ratio 15
1.3.6 Energy Saving and Environmental Protection 16

2 "CHSR Mode" Under the Innovative National Strategy 19
2.1 Innovation-Driven Development Strategy Program of China 19
2.1.1 Strategic Background 19
2.1.2 Strategic Requirements 21
2.1.3 Strategic Deployment 22
2.1.4 Strategic Tasks 23

2.2 Core Force in the Leap-Forward Development of CHSR 25
2.2.1 The Institutional and Administrative Forces Dominated by the Government 26
2.2.2 Market and Industrial Forces of Enterprises as Main Bodies 27
2.2.3 The Scientific and Technological and Talent Forces of Integrated Learning, Research and Production 28
2.2.4 The Mental Force of Dedication, Opening up and Innovation and the Cultural Force of Learning Widely from Others' Strong Points 29
2.2.5 Deep Integration of the Government, Enterprises and Universities as Main Innovation Bodies 29
2.3 HSR Experience in the Construction of Innovative Nation of China 30
2.3.1 Give Full Play to the Advantages of the Socialist System and Concentrate Forces on Major Tasks 30
2.3.2 Firmly Take the Road of Indigenous Innovation, and Maintain the Strategic Willpower of Taking Ourselves as the Main 32
2.3.3 Aim at "Exchange Market for Technology", to Fully Raise the Capability of Indigenous Innovation 32
2.3.4 Give Full Play to the Institutional Advantages and Main Force Role of State-Owned Enterprises in Important Fields and Key Sectors 33
2.4 The Global Strategic Value of China High-Speed Railway 33
2.4.1 Strategic Value of Regional Development: Re-Balance in the Development of the Eastern, Central and Western Regions 34
2.4.2 Strategic Value of Industrial Economy: Making the Updated Version of Chinese Economy 35
2.4.3 Strategic Value of International Economy: Shaping the Framework of Belt and Road International Economic Cooperation 36
2.4.4 Strategic Value of Diplomacy: Opening up the New Era of "HSR Diplomacy" 37
2.4.5 Strategic Value of Politics: Making the New Map of "Geopolitics" 38
2.4.6 Strategic Value of Finance: Reshaping the New World Financial Order 40

2.4.7 Strategic Value of National Defense: Reconstructing the "New Land Right", and Improving Remote Delivery Ability and Strategic Maneuverability 42
2.4.8 Strategic Value of Culture and Civilization: Building the "Community with a Shared Future for Mankind" 44

Part II The Belt and Road and CHSR "Going Global"

3 Belt and Road: Culture and Communications Aiming at Great Unity 49
3.1 Silk Road 50
3.1.1 Silk Road in Narrow Sense 50
3.1.2 Silk Road in Broad Sense 51
3.1.3 The Ancient Silk Road of China 52
3.2 The Belt Taking the Lead, and the Road Following 54
3.2.1 Arrangement for Belt and Road: Overall Planning for Land and Ocean, and Starting First on Land 54
3.2.2 The "Project Mausam" of India and Opening and Cooperation Development Strategies of Multiple Countries 56
3.2.3 Build the "Asian, European and African" Regional Economies on Both Land and Sea 58
3.3 Integration and Learning from Each Other for Inclusive Development—Extension of the Spirit of Ancient Silk Road in Modern Times 59
3.3.1 The Spirit of "Silk Road" 59
3.3.2 Social Development Is Originated from Trade Exchanges 60
3.3.3 Trade Exchanges Promote Cultural Integration 61
3.3.4 Cultural Integration Promotes People-to-People Bond 61
3.3.5 Build the "Commercial and Trade Circle" for Belt and Road with Road Connection as the Pilot 62
3.3.6 Forming the "Culture Circle" of "Belt and Road" with Mutual Progress as the Purpose 63
3.3.7 Developing the "Circle of Friends" of Belt and Road with the Aim of Inclusive Development 65
3.4 The Silk Road Calls for Development of Rail Transport 66
3.4.1 Rail Transport: The Forerunner for Belt and Road 66
3.4.2 Trans-Asian Railway: The Basic Framework of Belt and Road 66
3.4.3 Main Channels on Land: Urban Cluster Transportation Demand 67
3.5 "Big Transportation" in the Framework of Belt and Road 69

3.5.1 "Big Transportation" at the Physical and Technical Level: Connectivity of Land, Ocean and Air Infrastructures 69
3.5.2 "Big Transportation" at the Economic and Industrial Level: Integration of Infrastructure Construction and Industrialization and Urbanization 71
3.5.3 "Big Transportation" at the Culture and Civilization Level: Marching Towards the "Community with a Shared Future for Mankind" 73

4 Strategic Tension for China High-Speed Railway to Go Global 75
4.1 Five Tensions for CHSR to Go Global 75
4.1.1 Geopolitical Tension: Maritime Right and Land Right 75
4.1.2 Economic Tension: Demand and Effective Demand, Small Account and Big Account 77
4.1.3 Social Tension: Population Circulation and Conflicts Between Civilizations 80
4.1.4 Psychological Tension: Cognition and Identity 83
4.1.5 Tidal Tension: Globalization and Deglobalization 84
4.2 The CHSR Going Global: Work Hard to Press Ahead Despite Setbacks 88
4.2.1 Rare Opportunities 88
4.2.2 Variables Should not Be Underestimated 90
4.2.3 Achievements Are Accompanied by Setbacks 91
4.3 Ten Major Challenges for CHSR to Go Global 92
4.3.1 Rationally Cognize the Overseas HSR Market, Objectively Evaluate the Global HSR Capacity and View HSR Beyond It 93
4.3.2 Clearly Know the Diversified Conditions in Target Countries, and Precisely Grasp the Different Demands of Host Countries for HSR 94
4.3.3 Well Combine HSR Construction with Industrialization and Urbanization, to Upgrade the Industrialization and Urbanization with Railways 95
4.3.4 Pay Attention to Promotion of Value Identity, and Build the Community of Shared Interests, Responsibilities and Destiny 96
4.3.5 Gaming of Powerful HSR Countries and Breakout of Encirclement 99
4.3.6 It Is Imperative to Raise the International Recognition of China's HSR Standards 102
4.3.7 HSR Demands Huge Amount of Funds, and It Is Necessary to Break Through the Investment and Financing Bottleneck 105

4.3.8 It Is Imperative to Strengthen Top Level Design and Overall Planning . . . 108
4.3.9 Internationalized Inter-disciplinary Talents Are Short in China, and Professionals of Railway Technology Management Are Lacking in Countries Along the Route . . . 110
4.3.10 Multiple Realistic and Potential Risks Need to Be Resolved . . . 111

5 Strategy for CHSR to Go Global . . . 115
5.1 Strategic Orientations and Lines for CHSR to Go Global . . . 115
5.1.1 Strategic Directions: West, East, North and South . . . 115
5.1.2 Strategic Lines: Six Corridors . . . 116
5.2 Strategic Principles, Arrangement and Routes for CHSR to Go Global . . . 119
5.2.1 Strategic Principles . . . 119
5.2.2 Strategic Arrangement: Space Arrangement . . . 121
5.2.3 Strategic Arrangement: Time Echelon . . . 126
5.2.4 Strategic Routes . . . 127
5.2.5 "Accessing", "Entering", "Improving" and "Approaching" . . . 129
5.3 Strategic Main Points for CHSR to Go Global . . . 131
5.3.1 Establish and Improve Authoritative and Efficient Coordination Mechanism and Management Service Platforms . . . 131
5.3.2 Build an International Cutting-Edge HSR Industry System and a Continual Technology Innovation System . . . 133
5.3.3 Promote Internationalization of Chinese HSR Standards, and Build Brands of CHSR . . . 134
5.3.4 Speed up the Fostering and Training of Internationalized Talents for HSR . . . 136
5.3.5 Enhance the Risk Prevention and Control Systems . . . 137
5.3.6 Increase Financial and Banking Support . . . 139
5.3.7 From "Going Global" to "Entering Target Countries" . . . 141
5.4 Strategic Actions for CHSR to Go Global . . . 143
5.4.1 Set up a National Level "CHSR Going Global" Leading Group . . . 144
5.4.2 Establish the Portal Think Tank for CHSR Going Global . . . 145

5.4.3 Organize International Industrial Alliance for CHSR Going Global . . . 146
5.4.4 Create International Talent Education Basis for Rail Transport . . . 147
5.4.5 Construct "Rail Transport National Laboratory" . . . 148
5.4.6 Establish "Overseas Investment Venture Fund for High-Speed Rail Projects" . . . 149

Part III Undertakings of Universities in the High-Speed Railway Era

6 Higher Engineering Education in the Waves of "Reindustrialization" . . . 153
6.1 Engineering Science and Technology: The Power Motive Force for the Evolution of Human Civilization . . . 153
6.1.1 Outstanding Engineering Technology Achievements in Ancient China . . . 153
6.1.2 The Industrial Revolutions Have Demonstrated: Powerful Industries Means a Strong Nation . . . 153
6.2 Big Nation Manufacturing Industry Strategy in the Post-globalization Era . . . 154
6.2.1 The "Re-industrialization" in the United States . . . 155
6.2.2 "Industry 4.0" of Germany . . . 155
6.2.3 The "New Industries" and "Future Industries" of France . . . 156
6.2.4 The "New Growth Strategy" and "New Robot Strategy" of Japan . . . 156
6.2.5 Made in China 2025 Initiative . . . 156
6.3 Profound Transformation in Higher Engineering Education in World Important Economies . . . 158
6.3.1 The United States: The Washington Accord . . . 158
6.3.2 Germany: Craftsmanship . . . 159
6.3.3 The United Kingdom: The "Sandwich" Education Pattern and Professional Certification System . . . 159
6.3.4 France: Elite Engineers . . . 160
6.3.5 China: CDIO Engineering Education Pattern . . . 161
6.3.6 Featured Engineering Education Patterns in Japan, Korea and Australia . . . 161
6.4 Goals for Fostering Outstanding Engineering Talents . . . 162
6.4.1 Systematic, Broad and Solid Scientific and Technical Basic Knowledge and Engineering Professional Skills . . . 163
6.4.2 The Ability of Thinking to Become an Efficient Engineering Leader, Especially the Ability of Creative and Critical Thinking . . . 163

6.4.3 Proper Mastery of Digitalized Technology and Intelligent Tools ... 164
6.4.4 Healthy Personality, Professional Quality, Engineering Ethics and Humanistic and Artistic Accomplishment ... 164
6.4.5 The Comprehensive Quality of "Engineering with a Big E" ... 165

7 Future of High-Speed Railway and University Education ... 167
7.1 Transition of HSR Industry and Technological Leap-Forward in "Post-HSR Era" ... 167
7.1.1 The Change of China's High-Speed Rail Industry in the Post-ministry of Railways Era ... 167
7.1.2 Technical Leap-Forward in the "Post-HSR" Era ... 169
7.2 Mission and Journey of Research-Oriented Universities in the HSR Era ... 172
7.2.1 New Mission in the New Era: The Importance and Urgency of Fostering Innovative Talents ... 172
7.2.2 New Models and New Journey: The Challenges and Arduousness in Fostering Top-Notch Innovative Talents ... 174
7.3 Higher Education for Railway in China and Roles of Universities ... 175
7.3.1 Development Road of the Higher Education for Railway in China ... 175
7.3.2 Belt and Road and Undertaking by Universities and Colleges ... 177
7.4 Fully Support the HSR R&D and Construction: Exploration and Practices by Featured Research-Oriented Universities ... 179
7.4.1 Talent Fostering and Training ... 180
7.4.2 Scientific and Technological Innovation and Taking the Lead ... 180
7.4.3 International Exchange and Cooperation ... 182

Appendix ... 185

Postscript ... 201

About the Author

Dr. Fei Xu President of Southwest Jiaotong University, Professor of strategy, Ph.D. Supervisor, Senior Visiting Scholar of MIT and Harvard University; Vice President of China Railway Society, Vice Chairman of the Innovation and Entrepreneurship Society of China, Deputy Director of the Professional Board of Organization and Strategy of Chinese Management Society, Vice President of the Chinese Institute of Business Administration, Deputy Editor-in-Chief of *Strategic Management,* member of editorial board of *Systematic Management Journal* and *the Journal of the China Railway Society* . He published more than 10 books and translated works such as *Corporate Strategic Management*, *The Developing Theory and Growing Mechanism of the Firm*, *Theoretical and Empirical Study on Stability, Breaking, Evolution of Strategic Alliance*, *Dual Space and the Interduality of Management: An Exploration of the Fundamentals of Management*, *Systematics Interduality Theory: Theory and Methods*, *Profit Pattern 3.0: Competition and Cooperation in Time of Change*, *Great Minds in Management: the Process of Theory Development* (translated work), *Power of Culture* and completed more than 70 papers published in academic journals of home and abroad. He used to be the Executive Dean of Antai College of Economics & Management, Assistant President, and Vice President of Shanghai Jiaotong University. His main research fields are strategic management, game theory, innovative strategy, strategic alliance and competition strategy, and cross-culture leadership.

Part I
The Global Strategic Value of China High-Speed Railway

Since 2008, when the first High-Speed railway of China–Beijing–Tianjin Intercity Railway was completed for operation, China has developed its High-Speed railways most rapidly, with the most complete system technologies, the most powerful integration capability, the largest scale under construction, the longest operation mileage, the highest running speed, and the best performance-to-cost ratio in the world. China High-Speed Railway (CHSR), with the unremitting hard work by people of generations, has experienced the process from manufacturing to creation, from exploration to breakthrough and from catch-up to taking the lead in the course of independent development of trial trains, import to obtain mature experience on trains, and then self-reliance research and development of totally new trains on the basis of new experience. China High-Speed Railway has become a symbolic achievement in the construction of an innovative country together with manned spaceflight, manned deep-diving, lunar landing, supercomputers, and BeiDou satellites.

Chapter 1
China High-Speed Railway: Country's Golden Name Card

1.1 Three Climaxes of HSR Construction in the World

High-Speed railway, according to the definition of International Union of Railways (UIC), normally refers to railways with design speed over 250 km/h for new lines, or those with speed over 200 km/h for those with increased speed rating. In China, the definition for HSR is slightly different, and it specifically refers to newly constructed dedicated passenger transport railways for motor train units (MTU) with running speed of 250 km/h (including that reserved for the future), and initial period operation speed no less than 200 km/h. The tracking interval of HSR trains is designed as minimum of 3 m, with car axle load no more than 17t and no more than 16 cars in a train.[1] Globally, the development of HSR has experienced three climaxes of construction.[2]

1.1.1 The First Wave (From the 1960s to the End of 1980s)

The first wave was started in the 1960s, and ended in the late 1980s. Some developed countries, such as Japan, France, Italy and Germany, built their own High-Speed rail lines. The representative ones included the Shinkansen in Japan, the TGV lines in the southeast and along the Atlantic coast of France, the Roma–Florence Line in Italy, and Hanover–Wurzburg High-Speed new line in Germany. The total mileage of HSR in the world reached 3198 km.

[1] Refer to *Main Technology Policies for Railway* Rev. 6 issued by the former Ministry of Railways on January 9, 2013.

[2] He Shang: *Third Wave of World Railway Development, China Report* No. 12, 2010, pp. 46–47.

F. Xu, *The Belt and Road*, https://doi.org/10.1007/978-981-13-1105-5_1

1.1.2 *The Second Wave (From the End of 1980s to the Mid-1990s)*

The second wave was started from the end of 1980s, and continued to the mid-1990s. Due to the demonstration effect of HSR construction in Japan and other countries, various countries in the world paid great attention to it and put it into practice. In Europe, France, Germany, Italy, Spain, Belgium, Holland, Sweden and the UK were the most prominent performers: Sweden put into service the X2000 "tilting train" in 1991; in 1992, Spain imported technologies from France and Germany to build the 471 km long Madrid–Seville High-Speed railway. The first High-Speed rail international line connected France with Britain via the Channel Tunnel in 1994; and in 1997, the "Eurostar" train from Paris connected France, Belgium, Holland and Germany.

1.1.3 *The Third Wave (From the Mid-1990s Up to Now)*

The third wave has continued from the mid-1990s up to now, covering Asia, Europe, North America and Oceania, really a revolution in the world's transport industry. Countries and regions including Russia, South Korea, Australia, Britain, Holland and Taiwan of China started High-Speed railway construction one after another. To cooperate with the construction of HSR network in Europe, Czech, Hungary, Poland, Austria, Greece and Romania in Eastern and Central Europe also did a comprehensive reconstruction of their main railway lines to increase the speed. In addition, the United States, Canada, India and Turkey also began to pay attention to High-Speed railway.

In this third wave, CHSR surpassed the formers as a latecomer, with not only advanced technologies, safe and reliable performance, but also good compatibility and high performance to cost ratio.

1.2 Overtaking on Curved Lane: Innovation History of CHSR

The development of CHSR has experienced a process of starting from scratch, growing from weak to strong, and from "following the others" to "running side by side" and then "taking the lead", realized the leap-forward surpassing from "technology import" to "made in China" and then to "created in China", and has established a complete system of survey, design, construction, equipment, operation and safety management standards for HSR, the brands and proprietary intellectual property rights for HSR equipment, successfully opening up a road of leap-forward indigenous innovation in HSR with Chinese characteristics. CHSR has become a benchmark of "made in China" and "Chinese speed".

The innovation course of CHSR can be roughly divided into four periods: technology accumulation period, the period for technology import, digestion, absorption and re-innovation, the indigenous innovation period, and all-round indigenous innovation period (Fig. 1.1).

1.2.1 Technology Accumulation Period (Before 2004)

This period laid a solid foundation for the innovative development of CHSR. Early in the beginning of 1990s, China started technological research and engineering practices on HSR. At that time, the State Commission of Science and Technology included the key scientific and technological project "complete set of HSR technologies" into the state key scientific and technological program of the 8th Five-Year Plan period, organized relevant scientific research forces to work jointly on key technical issues in HSR in China, making the necessary initial technical reservation for the subsequent construction of Beijing–Shanghai High-Speed Railway and 6 railway speed raise projects. The large area speed increase projects of existing railways since 1997, and the Qinhuangdao–Shenyang HSR which was started for construction in August 1998 and put into operation in October 2003 accumulated experience for the development of HSR in China.

In this period, the Chinese railway sector, by relying on the own technical forces and indigenous innovation, developed more than 10 types of High Speed trains represented by "China Star". In this period, although there was a fairly big gap between the High-Speed trains developed in China and those in advanced foreign countries, China laid the initial technological development basis for HSR manufacturers, form-

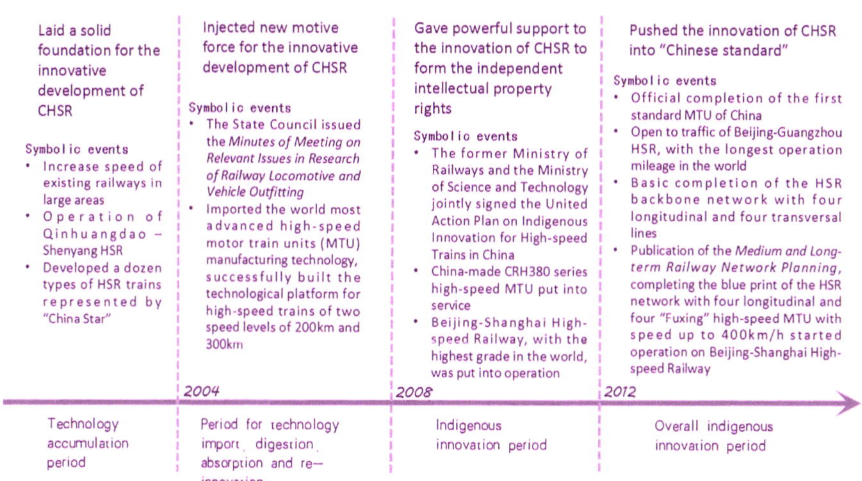

Fig. 1.1 Innovation course of CHSR

ing the initial HSR technical system, fostered and tempered the team of talents, and provided the technology and talent reservation for the technology import, digestion, absorption and re-innovation in the subsequent period.

1.2.2 Period for Technology Import, Digestion, Absorption and Re-innovation (2004–2008)

This period injected new motive force for the innovative development of CHSR. In April 2004, the State Council issued the *Minutes of Meeting on Relevant Issues in Research of Railway Locomotive and Vehicle Outfitting*, making clear the technical lines for the import, digestion, absorption and re-innovation of High-Speed motor train units. The former Ministry of Railway and National Development and Reform Commission jointly issued the *Implementation Plan for Technology Import and Localization of Motor Train Units of Hourly Speed of 200 km and Over*, deciding to implement the action plan "with overall technology transfer as the prerequisite, the import of core and key technologies as the main point, the domestic enterprises as the main bodies, and localization as the ultimate goal".

Specifically, China would concentrate forces to import the world most advanced High-Speed motor train unit (MTU) manufacturing technology, to successfully build the technological platform for High-Speed trains of two speed levels of 200 and 300 km; China would pool resources of all aspects, and organize the relevant rail transport sector in China, to carry out digestion, absorption and re-innovation of imported technologies, to master nine core technologies and ten main associated technologies of High-Speed MTU. This process injected new motive force for the innovative development of CHSR, greatly pushed ahead the overall progress of CHSR in basic theory and research of key technologies, substantially upgraded the technical outfitting level of CHSR, and also accelerated the penetration of relevant technologies into the industrial chain of CHSR while rapidly upgrading the HSR technologies.

1.2.3 Indigenous Innovation Period (2008–2012)

This period gave powerful support to the innovation of CHSR to form the proprietary intellectual property rights. In February 2008, the former Ministry of Railways and the Ministry of Science and Technology jointly signed the *United Action Plan on Indigenous Innovation for High-Speed Trains in China*, and it stated to establish and complete the CHSR technology system with proprietary intellectual property rights and powerful international competing power of 350 km/h and over on the basis of digesting and absorbing advanced technologies from foreign countries, to meet the operation demand of Beijing–Shanghai Express Railway at a speed of 380 km/h. The united actions of the two ministries were supported by the project "development of key technologies and equipment for High-Speed trains in China" in the 11th Five-Year Plan national scientific and technological support program, to realize scientific and technological innovation in the form of the government taking the lead and enterprises acting as the main body, with market as orientation and project cooperation as ribbon, and push forward the positive design and indigenous innovation for High-Speed trains in China.

In August 2008, the first railway with design speed of 350 km/h and running over soft soil area in China, Beijing–Tianjin Express Railway, was put into operation. In 2010, the China-made CRH380 series High-Speed MTU, with high degree of proprietary intellectual property rights, was put into service, successfully realizing breakthroughs in key technologies of head-to-head, lightweight vehicle body, bogie, vibration and noise reduction and system integration, laying a solid foundation for realizing overall indigenous innovation of core technologies and pushing the CHSR into the "Chinese standard" phase in the next period. In June 2011, Beijing–Shanghai Express Railway, with the highest grade in the world and able to operate at the highest speed of 486 km/h (with a total length of 1318 km), was put into operation.

1.2.4 All-Round Indigenous Innovation Period (Since 2012)

This period pushed the innovation of CHSR into "Chinese standard". In 2012, according to the innovation-driven development strategy of the state, we actively pushed ahead the development of MTU to Chinese standards with full proprietary intellectual property rights in the principle of adhering to proprietary intellectual property rights, safety and reliability, standardization, serialization, simplicity and unification, economic performance, energy saving and environmental protection. In June 2015, the first standard MTU of China was officially completed, marking that the innovation of CHSR officially has stepped into the era of "Chinese standards".

The technical standard system for Chinese MTU, while adopting many national standards, industrial standards and special technical standards of China, has also actively integrated with some international standards and advanced technical standards of foreign countries. Also, with the export of High-Speed MTU as breakthrough, the export of complete set of HSR technologies for construction engineering, traction and power supply, communication signal and operation management was started, to build the "CHSR" brand and raise the competing power of CHSR on the international market on all aspects, to realize the strategic goal of CHSR "going global".

In December 2012, Harbin–Dalian HSR, the first HSR line running through high and cold area with seasonal frozen soil in the world, was completed for operation; also in this month, Beijing–Guangzhou HSR, with the longest mileage in the world (with a total length of 2298 km) and covering many topographical and geological areas in temperate and subtropical zones and a number of river system, was open to traffic the whole line; in December 2014, Lanzhou–Urumqi HSR, with the longest mileage completed one time in the world (1777 km) and running through desert and strong wind zones, was put into operation; in December 2015, the ring HSR in Hainan, the first island ring HSR in the world with a length of 653 km, and also the HSR at the most south part in the world up to now, was open for operation.

By the end of 2015, China basically completed the HSR trunk network with four transversal and four longitudinal lines, and its HSR operation mileage reached 19,000 km, ranking the first in the world, and accounting for over 60% of the world total HSR mileage, including 9661 km of HSR with speed of 300 km/h. According to the 13th Five-Year Plan of the state, China will speed up the completion of its HSR network in the coming five years, and by 2020, its HSR operation mileage will reach 30,000 km, linking over 80% of its big cities. In 2015 China also reached agreements with Russia and Indonesia, to build in cooperation the Moscow–Kazan HSR and Jakarta–Bandung HSR, successfully realizing the breakthrough progress for the CHSR to go from domestic to the world market.

In July 2016, the National Development and Reform Commission, Ministry of Transport and China Railway Corporation jointly published the *Medium and Long-term Railway Network Planning*, as a grand blueprint of the HSR network with "eight longitudinal and eight transversal lines in the new period (Table 1.1 and Fig. 1.2). The "eight longitudinal lines" include: the coastal line, Beijing–Shanghai, Beijing–Hong Kong (Taipei), Beijing and Harbin–Beijing Hong Kong and Macau, Hohhot–Nanning, Beijing–Kunming, Baotou (Yinchuan)–Hainan, and Lanzhou (Xining)–Guangzhou. The "eight transversal lines" include: Suifenhe–Manzhouli, Beijing–Lanzhou, Qingdao–Yinchuan, the Continental Bridge Line, the line along the Yangtze River, Shanghai–Kunming, Xiamen–Chongqing, and Guangzhou–Kunming lines. This HSR network with eight longitudinal and eight transversal lines will include the eight longitudinal lines including the coastal and Beijing–Shanghai lines and the eight transversal lines including the Continental Bridge Line, the line along the Yangtze River as the main, and supplemented by inter-city rails. This network with eight longitudinal and eight transversal lines will realize traffic circles of 1–4 h between adjacent big and medium sized cities and 0.5–2 h within city clusters, connecting all big and medium sized cities with population of over 500,000 each. By 2025, the HSR operation mileage will reach 38,000 km.

Through continuous indigenous innovation, major breakthrough has been achieved in a number of core key technologies in the field of HSR vehicle and equipment manufacturing: The core technologies such as traction and current inversion, microcomputer network technology, and braking technology have broken the technological and industrial monopoly by foreign countries; the development and development of a series of advanced, safe, reliable and price competitive products of high-end rail transportation represented by HSR have been developed; fairly complete innovation platform, product development platforms and manufacturing platforms for rail transportation equipment have been established; the MTU successfully developed to Chinese standards has embodied the national strategy to achieve breakthrough in MTU core technologies and master the proprietary intellectual property rights.

On June 25, 2017, the MTU to Chinese standards developed with China Railway Corporation taking the lead, with full proprietary intellectual property rights

Table 1.1 The main planned "eight longitudinal and eight transversal" lines of HSR in China

Eight longitudinal lines	The coastal line	Dalian (Dandong)–Qinhuangdao–Tianjin–Dongying–Weifang–Qingdao (Yantai)–Lianyungang–Yancheng–Nantong–Shanghai–Ningbo–Fuzhou–Xiamen–Shenzhen –Zhanjiang–Beihai (Fangchenggang)
	Beijing–Shanghai line	Beijing–Tianjin–Jinan–Nanjing–Shanghai (Hangzhou), Nanjing–Hangzhou, Bengbu–Hefei–Hangzhou, Beijing–Tianjin–Dongying–Weifang–Linyi–Huai'an–Yangzhou–Nantong
	Beijing–Hong Kong (Taipei) line	Beijing–Hengshui–Heze–Shangqiu–Fuyang–Hefei (Huanggang)–Jiujiang–Nanchang–Ganzhou–Shenzhen–Hong Kong (Kowloon), Hefei–Fuzhou–Taipei, Nanchang–Fuzhou (Putian)
	Beijing and Harbin–Beijing Hong Kong and Macau lines	Harbin–Changchun–Shenyang–Beijing–Shijiazhuang–Zhengzhou–Wuhan–Changsha –Guangzhou–Shenzhen–Hong Kong, Guangzhou–Zhuhai–Macau
	Hohhot–Nanning line	Hohhot–Datong–Taiyuan–Changzhi–Jincheng–Jiaozuo–Zhengzhou–Xiangyang–Changde–Yiyang –Loudi–Shaoyang–Yongzhou–Guilin–Nanning
	Beijing–Kunming line	Beijing–Shijiazhuang–Taiyuan–Xi'an–Chengdu (Chongqing)–Kunming line, Beijing–Zhangjiakou–Datong–Taiyuan
	Baotou (Yinchuan)–Hainan line	Baotou–Yan'an–Xi'an–Chongqing–Guiyang–Nanning–Zhanjiang–Haikou (Sanya), Yinchuan–Xi'an, Hainan Island Ring Rail
	Lanzhou (Xining)–Guangzhou line	Lanzhou (Xining)–Chengdu (Chongqing)–Guiyang–Guangzhou

(continued)

Table 1.1 (continued)

The eight transversal lines	Suifenhe–Manzhouli line	Suifenhe–Mudanjiang–Harbin–Qiqihar–Hailar–Manzhouli
	Beijing–Lanzhou line	Beijing–Hohhot–Yinchuan–Lanzhou
	Qingdao–Yinchuan line	Qingdao–Jinan–Shijiazhuang–Taiyuan–Yinchuan
	Continental Bridge Line	Lianyungang–Xuzhou–Zhengzhou–Xi'an–Lanzhou–Xining–Urumqi
	The line along the Yangtze River	Shanghai–Nanjing–Hefei–Wuhan–Chongqing–Chengdu, Nanjing–Anqing–Jiujiang–Wuhan–Yichang–Chongqing, Wanzhou–Dazhou–Suining–Chengdu
	Shanghai–Kunming line	Shanghai–Hangzhou–Nanchang–Changsha–Guiyang–Kunming
	Xiamen–Chongqing line	Xiamen–Longyan–Ganzhou–Changsha–Changde–Zhangjiajie–Qianjiang–Chongqing
	Guangzhou–Kunming line	Guangzhou–Nanning–Kunming

Source National Development and Reform Commission, Ministry of Transport, China Railway Corporation, *Medium and Long-term Railway Network Planning*, July 2016

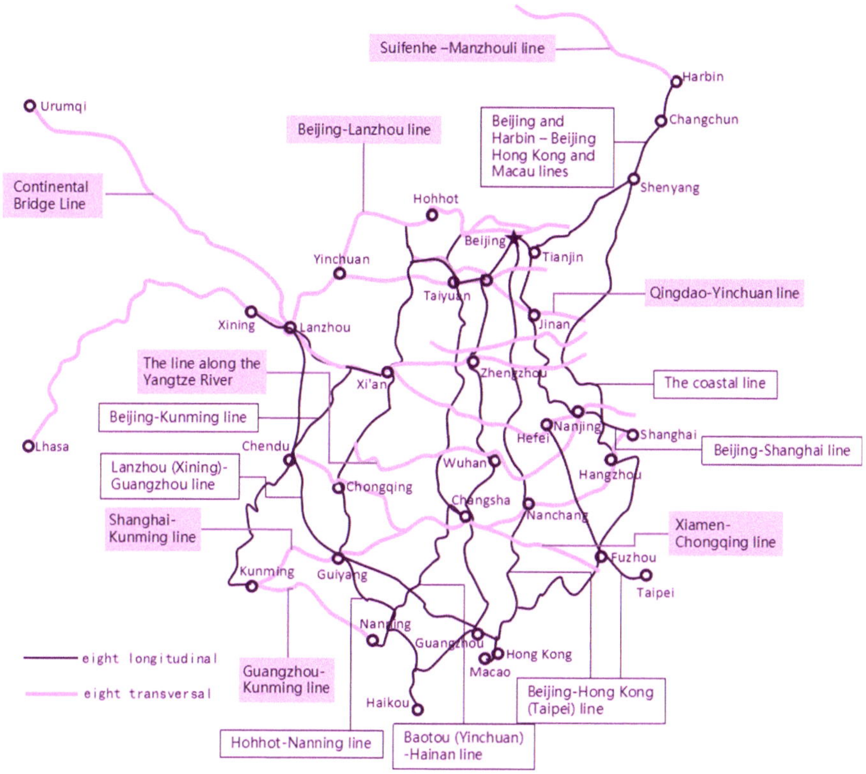

Fig. 1.2 Schematic diagram of planned "eight longitudinal and eight transversal" lines of HSR network in China

and meeting the world advanced level was named "Fuxing", and its speed can reach 400 km/h. This standard MTU is now available in two models: "CR400AF" and "CR400BF". On June 26, "Fuxing" set out the first time from both ends of Beijing–Shanghai High-Speed Railway, the Beijing South Station and Shanghai Hongqiao Station. Subsequently, China will, on the basis of the platform of "Fuxing" MTU, develop independent and standardized MTU series products for different speed levels and suitable to different environment demands with continual scientific and technological innovation on aspects of safety, economy, comfort, energy saving and humanization.

CHSR has obtained great achievements, won extensive praises in all walks of life including the international peers, and fully confirmed by the leaders of the CPC Central Committee and State Council. After inspection of relevant enterprises, General Secretary Xi Jinping pointed out that, "HSR is a bright name card of equipment manufacturing of China". Premier Li Keqiang, as a "super salesman" for HSR, emphasized at different occasions that CHSR has the great advantages of "advanced technologies, good safety and reliability, high performance to price ratio and rich operation experience", and said a number of times that "I'm particularly emboldened in promoting CHSR". It is perhaps a rare thing in the reform and opening up of China that the premier is doing his best to promote the sale of HSR overseas.

At present, CHSR tightly grasps the rare opportunities that the state is implementing the Belt and Road Initiative, energetically implements the internationalized operation strategy, and actively opens up overseas operation businesses, and its brand value and international influence has been increasing year by year. The modern rail transportation equipment represented by HSR has become the pace-setter for the high-end equipment of China to go global. By now, the CHSR products have been exported to nearly 100 countries and regions in the six continents all over the world, the exported products have realized upgrading from medium and low end to high end, the export market realized a leap-forward from Asia–Africa–Latin America to Europe and America, and multiple forms of export have been realized from sole product to the combined export of products, capital, technology and services.

1.3 Achievements of CHSR that Attract World Attention

We see in a recall of the world HSR development history that, Japan realized operation of HSR with 2300 km in half a century, with average operation speed of 243 km/h; France built over 1900 km of TGV HSR in more than 40 years, with average operation speed of 277 km/h; and Germany built nearly 1600 km of ICE HSR in over 20 years with average operation speed of 232 km/h. China built and put into service 19,000 km of HSR in less than 7 years, exceeding the total of operating HSR in all other countries in the world.[3] The achievements of China in HSR have attracted the world attention, and they are described in detail below.

1.3.1 The Largest Scale

By the end of 2016, the operation mileage of CHSR had exceeded 22,000 km, making China the country in the world with the longest HSR operation mileage, largest scale under construction, the largest number of MTU trains and the busiest operation.

Taking the 2015 National Day "golden week" as an example, the whole country put 3672 MTU trains in operation, 852 trains more than that of the same period of previous year, they transported total of 43.407 million passenger-times, up 30.3% from the same period of previous year, or 4.341 million passenger-times by daily average, accounting for 44.4% of total railway passenger transport volume of the whole country. In September 30 alone, 4.809 million passengers were transported, set a highest record of passenger transport by MTU in a single day. The passenger transport vol-

[3]Sheng Guangzhu: *China High-Speed Railway Advancing in Big Strides, Truth Seeking* No. 19, 2014, pp. 31–33.

ume of nearly 20 HSR lines including Beijing–Shanghai and Beijing–Guangzhou lines set new records in history, with the highest daily passenger transport volume being 430,000 with Beijing–Shanghai HSR and 492,000 with Beijing–Guangzhou HSR. This shows the development momentum of CHSR.

1.3.2 Advanced Technologies

In project construction, in order to adapt to the complex and varied geological and climate conditions in China, a series of difficult technical issues have been solved in the infrastructure construction for the roadbed, bridges, tunnels and terminals, and in professional fields of ballastless track, traction power supply, and communication signals.

In High-Speed MTU, China developed key technologies and associated technologies for it through import, digestion, absorption and re-innovation, basically formed the technical standards system for two series of MTU's of 200–250 km/h and 300–350 km/h, and realized domestic design and manufacturing, creating the brand of MTU to Chinese standards.

On December 3, 2010, the CRH380AL High-Speed MTU reached the highest testing speed of 486.1 km/h in Zaozhuang–Bengbu Section of Beijing–Shanghai HSR, refreshing again the highest record of 416.6 km/h on operating railway in the world previously set in Shanghai–Hangzhou HSR. It should be pointed out that early on April 3, 2007, the High-Speed train V150 of France reached 574.8 km/h in a test run, however, the highest speed of 486.1 km/h in China was realized with MTU CRH380A in normal operation on routine operation lines, and after the operation, the train still remain intact without damage. But the train in France setting the highest speed of 574.8 km/h was specially retrofitted for test, the locomotive was of narrow body with 4 seats in a row, and the operation line was also specially built with granite special grade ballast. After the end of test, the train was almost totally scrapped.

In operation management, China has mastered the HSR operation dispatching technology under complicated network conditions, established the passenger service system suitable to high passenger flow with high density, and built the HSR safety risk prevention and control system, to provide reliable technical guarantee for the safe operation of HSR.

In the future, CHSR will develop traction power supply system with no split-phase and with remote monitoring, the train control system based on LTE communication and the stationary and mobile equipment smart monitoring and pre-warning technologies based on big data, to secure its leading position in HSR in the world.[4]

---

[4]Lu Chunfang: *Technical Features of China High-Speed Railway, Science & Technology Review* No. 18, 2015, pp. 13–19.

1.3.3 Safe and Reliable

Needless to say, there has been misunderstanding in the public opinion concerning the safety of CHSR since the July 23 incident on Ningbo–Wenzhou Line in 2011, and the public opinion "shadow" has not been completely eliminated even today. In fact, the train speed when the accident occurred was only about 100 km/h, not related to the High-Speed of railway, and the cause of the accident was neither track nor train, it was only the problem in the operation and commanding signal system.

At present, the design running speed of CHSR can reach 400 km/h. It is not out of worry about safety that the trains are not running at the highest rated speed. "According to the current rail technology of China, we can absolutely guarantee safety even if we run it at 385 km/h. We seek for the most "economic" practical running speed only out of more consideration on extending the service life of trains and tracks."[5]

In fact, for the operation of CHSR, not only safety guarantee system with closed-loop management has been built, to realize the real-time acquisition and analysis of messages from various mobile devices and stationary facilities, a massive railway dispatching and commanding system has also been built, to powerfully ensure high density operation of trains. At present, the whole railway system runs over 7000 passenger trains every day, including over 3000 MTU's, and over 20,000 freight trains.

More than that, verification is made before opening for traffic with the advanced High Speed comprehensive testing train, and after operation, all lines are inspected in turn every 10 days. Furthermore, the seismic monitoring and pre-warning system is being developed for the national seismic monitoring stations and network, to release the pre-warning messages of earthquake intensity and arrival time by using the speed difference of electromagnetic wave and seismic wave and the speed difference between P wave and S wave, so that pre-warning of several to several dozen seconds in advance can be provided for HSR in the affected area, to enable braking of High-Speed train in advance.

1.3.4 Good Compatibility

CHSR is well compatible with advanced technologies in the world in the main areas of project construction, MTU, train control and traction power supply. It has not only integrated with advanced international standards such as UIC (standards of International Union of Railways), IEC (International Electrotechnical Commission standards), ISO (International Standardization Organization standards), EN (European standards), and JIS (Japan Industrial Standards), but also is fully compatible with

[5]Wang Mengshu: *High-Speed Railway is too Fast and not Safe? Absolutely No Problem even at 385 km/h*, *The Paper*, October 17, 2015.

Siemens of Germany (Velaro-E), Kawasaki Heavy Industries of Japan (E2-1000), Alstom of France (SM3) and Bombardier of Canada (Regina).

Good compatibility is originated from the outstanding system integration and innovation capability in the development of CHSR. Today, China has not only fully mastered the critical technologies of overall design, interface management and combined commissioning and testing for HSR, but also is carrying out research on cutting-edge issues to further lower the full life-cycle cost of HSR operation, increase the train dispatching efficiency and mitigate the aging and wearing of parts of trains with the backup by the big data of CHSR operation, to continually optimize the overall performance of HSR.[6]

1.3.5 High Performance to Price Ratio

The high performance to price ratio of CHSR is first embodied in the construction schedule and quality. The optimal match of construction plan, resources allocation and control targets has been realized by innovating the dynamic management pattern in construction organization and with factory fabrication and mechanization as support, thus greatly increasing the construction efficiency and ensuring the schedule and quality. Short work schedule is realized by scientific calculation and reasonable determination of schedule, instead of unreasonably shortening schedule. Therefore, the construction period of CHSR is much shorter than that of other countries (Fig. 1.3).

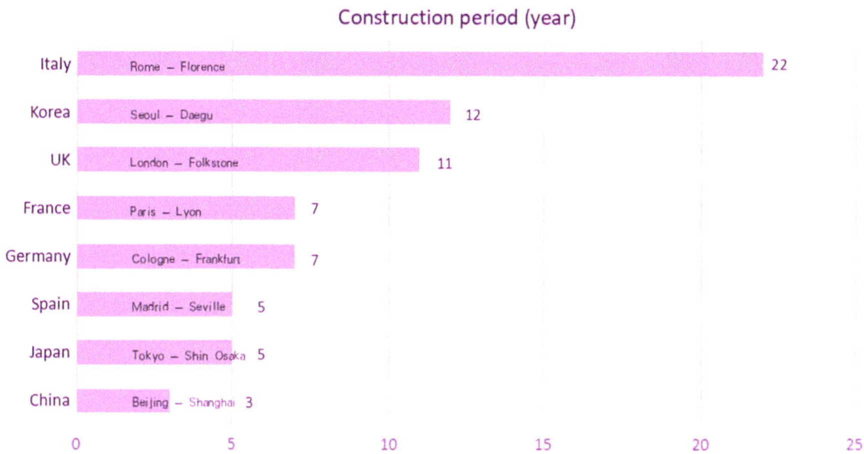

Fig. 1.3 Comparison of HSR construction periods in different countries in the world

[6] *Where are the Strong Points of China High-Speed Railway, Xinmin Weekly* No. 41, 2015, pp. 6–7.

Secondly, according to the World Bank research report on July 2014, the per kilometer construction cost of CHSR is about 2/3 that of developed countries. For newly developed trains, advanced light materials such as magnesium alloy and carbon fiber are used, the low resistance design with "Chinese elements" and high efficiency traction and braking systems are used, and attention is paid to parts most likely to wear, so that the vehicle life can be 30 years.

1.3.6 Energy Saving and Environmental Protection

Energy saving and environmental protection is one of the major advantages of CHSR. Researches show that, suppose the energy consumption per person per kilometer is 1.0 with ordinary railway, it is 1.42 with HSR, 8.5 with cars and 7.44 with aircraft. More than that, with the progress of energy saving technologies for HSR, continual progress is being made in its energy saving effect. The UIC research has indicated that, on the same line, High-Speed trains consume less energy than ordinary trains.[7] At present, the per capita power consumption per 100 km by High-Speed MTU in China is less than 8 kWh.[8]

For HSR stations, solar energy PV power generation, ground source heat pump and other new energy technologies are used. In land saving and noise reduction, bridges are built in large numbers in place of subgrade on ground. Compared with subgrade, bridges can save land by 3/5. During construction, water and soil conservation measures such as side slope vegetation protection and soil covering for re-cultivation and plantation are adopted for subgrade, and sound barriers and vibration reduction provisions are arranged, to effectively reduce the impact of HSR noise to the environment.

It is worth mentioning that China is one of the very few countries in the world that has mastered the permanent magnetic traction system technology for HSR. In terms of evolution of technical systems, from DC drive traction system to AC drive traction system, and from induction asynchronous drive to permanent synchronous drive, the permanent synchronous traction system is in line with the current trend of technological development for energy conservation and emission reduction and green environmental protection, so it has become a hot technical point researched by all major countries in the world. The permanent synchronous traction system developed by China has demonstrated the advantages of high efficiency and high power density, and can substantially reduce the traction energy consumption for High-Speed trains.

---

[7]Zhou Xinjun: *China Needs to Strive to Develop High-Speed Railway—Also on the Energy Conservation and Emission Reduction Effect of High-Speed Railway, China Economic Report* No. 7, 2015, pp. 66–69.

[8]Lu Chunfang: *Technical Features of China High-Speed Railway, Science & Technology Review* No. 18, 2015, pp. 13–19.

At present, the rated efficiency of motor has reached over 98%, and the motor loss has been reduced to 1/3 of the original level. In the great background of the era that all countries in the world pursue for "green traffic", permanent synchronous traction system has become a model of the HSR technology in China.[9]

[9]Cao Chang and Li Yonghua: *China High-Speed Railway Adopts the World Most Advanced Traction Technology*, *China Economic Weekly* No. 24, 2015, pp. 43–45.

Chapter 2
"CHSR Mode" Under the Innovative National Strategy

2.1 Innovation-Driven Development Strategy Program of China

The 18th CPC National Congress proposed the strategy of innovation-driven development and stressed that scientific and technological innovation is a strategic pillar for boosting social productivity and comprehensive national strength and therefore must be placed at the heart of the country's development. The Fifth Plenum of the 18th CPC Central Committee put forth the development concept of "innovation, coordination, green, opening and sharing", and the base of development is placed on innovation, which becomes the first motive force that leads the development. In 2016, the Central Committee of CPC and State Council published the *National Innovation-driven Development Strategy Outline* (hereinafter referred to as the Outline), which is a major development strategy of the state determined by the CPC Central Committee in the new development phase with a global view, focusing on critical issues and to drive up the entirety.[1]

2.1.1 Strategic Background

Innovation-driven development means innovation should become the primary driver of development, that innovation in science and technology must be combined with innovation in system, management, business model, and culture, and that efforts must be made to pursue development on the basis of continued knowledge accumulation, technological progress and the improvement in labor competence, and

[1] *Central Committee of CPC and State Council Printed and Issued "National Innovation-driven Development Strategy Outline", People's Daily* May 20, 2016, p. 1.

F. Xu, *The Belt and Road*, https://doi.org/10.1007/978-981-13-1105-5_2

promote economic growth with more advanced form, more sophisticated division of labor and more efficient structure

The world today is in a grand transformation and readjustment. The global economic growth has slowed down, and raising total factor productivity to gain new growth point has become the strategic choice for all countries in the world to seek a new round of economic prosperity. At present, a new round of S&T revolution and industrial evolution is picking up speed on the global stage, scientific exploration is deepening from microscopic to cosmic scales, group technological revolution characterized with intelligence, green, ubiquity and sustainability will trigger a major adjustment of international industrial division of labor, subversive technologies are constantly emerging, reshaping the world competition pattern, and changing the relative strength of nations. Innovation drive has become the core strategy of many countries seeking for competitive advantage.

In China today, science and technology innovation has become the critical forces and sharp weapons to support the national development and safeguard the national security, playing the role as an engine in modernization construction and to realize the "two centenary goals". After decades of unremitting efforts, Chinese sciences and technologies have achieved a large number of results in many fields with international influence. The continuous breakthroughs in manned spaceflight, manned deep-diving, lunar exploration, supercomputing, Beidou satellite and artificial intelligence have led to the development of related science, technology and engineering fields. Technologies are becoming daily mature in fields of High-Speed railway, bridge, tunnel, UHV power transmission and transformation, new energy (wind power, nuclear power, photovoltaic), highly difficult oil and gas fields, super rice, mobile Internet, and mobile payment, and are being exported abroad.

In a number of leading and emerging field of research such as iron-based superconductivity, neutrinos, quantum information, Weyl fermion, nanotechnology, space science, stem cell and regenerative medicine, origins and evolution of life and so on, China has also made a series of world leading major achievements.[2] According to the *2016 Research Frontiers* report published by institutions such as Institute of Science and Development, Chinese Academy of Sciences, of the 180 international frontier and emerging frontiers, China performs excellently in 30 research frontiers, exceeding the UK, Germany, Japan, France, ranking at the second place in the world, only next to the United States, mainstream disciplines such as chemistry, materials, physics, engineering, mathematics, and geography are approaching the top level of the world.

While fully affirming the rapid rise of China's science and technology, it is necessary to deeply recognize the existing gap of China in scientific and technological innovation. In the key stage to realize the Chinese Dream for great rejuvenation of the Chinese nation and the critical period of innovation driven transition development, we can get the initiative to develop only by bravely holding on the forefront of science and technology innovation, to make greater contribution to human civilization

[2]Pan Jiaofeng: *China Accelerates towards a World Innovation Center, Reference News* March 16, 2017, p. 11.

and progress. In the future, grasping innovation means development, and seeking for innovation is making the future. Innovation will be made as a national will and the common action of the whole society, and we must open a new road of development from powerful talents and science and technology to powerful industries, economy and nation, so as to create a new growth period for the future of China for dozens years or a longer.

2.1.2 Strategic Requirements

We should adhere to the road of indigenous innovation with Chinese characteristics, emancipate the mind, be open and inclusive, take innovation-driven development as the priority strategy of the country, drive the comprehensive innovation with science and technology innovation as the core, inspire creativity with reform of systems and mechanisms, support the construction of a high level innovation-oriented country with highly efficient innovation system, and promote the fundamental transformation of economic and social development motive force, to provide powerful motive force for the realization of the Chinese Dream for the great rejuvenation of the Chinese nation.

The *Outline* put forth the following basic principles:

Grasp the development. We should adhere to orientation by problems, face the cutting edge of world science and technology, the major demand of the state and the main battlefield of national economy, make clear the main direction of innovative development of China, realize breakthrough in key areas, and strive to gain more competition advantages.

Deepen the reform. We should make efforts concurrently in reform of scientific and technological system and economic and social fields, better connect science and technology with economy, observe the economic law and scientific and technological innovation law on socialist market, break all ideological barriers and institutional fences restricting innovation, and build up a good environment supporting the innovation-driven development.

Enhance incentives. The essence of innovation drive is talent drive, we should take people as the fundamental, respect the value created by innovation, stimulate the initiative and creativity of all categories of talents, and quickly gather a team of innovation-oriented talents with large scale, reasonable structure and excellent quality.

Expand opening-up. We should plan and promote innovation with a global vision, make the maximum use of global innovation resources, fully raise the potential of China in the global innovation pattern, and strive to become the leader in some important fields and the participator in formulating important rules.

The *Outline* clearly sets the three-step strategic goal: as the first step, China will rank itself among the innovative countries by 2020, basically complete the construction of a national innovation system with Chinese characteristics, to powerfully support the realization of the goal of completing the building of a moderately prosperous society.

As the second step, by 2030, China will rank at the front of innovative countries, realize a fundamental change in the drive force for development, the economic and social development level and the international competing power will increase substantially, to lay a solid foundation for building a powerful nation of economy and a society of common prosperity.

As the third step, by 2050, China will become a powerful nation of scientific and technological innovation in the world, and a main scientific center and innovation highland in the world, to provide powerful support to building China into a great modern socialist country that is prosperous, strong, democratic, culturally advanced and harmonious and realizing the Chinese Dream of national rejuvenation.

2.1.3 Strategic Deployment

Realizing innovation drive is a major strategic choice of the CPC Central Committee and State Council facing the world, the future and the modernization. Innovation drive is a systematic transformation, and we should make arrangement according to "driving by twin wheels, building one system and promoting six transformations", to build a new motive force system for development.

Driving by twin wheels means the two wheels of STI and institutional innovation should coordinate with each other and play their role sustainably. Innovation is, first and foremost, about innovation in science and technology. It is imperative to strengthen the weak links in STI before anything else. We must be clear about the direction and priority of development, strengthen scientific exploration and make technological breakthroughs, and foster a systemic capability for sustained innovation. Institutional innovation should aim at adjusting all production relations inconsistent with the goal of innovation-driven development. We should advance institutional innovation concerning science and technology, economy and government administration, unleashing fully the power of innovation.

One system means the building of the national innovation system. We should build an ecosystem featuring close coordination and interaction among various innovation entities as well as unimpeded flow and efficient allocation of innovation factors, and bring about the carriers, institutional arrangements and safeguards for achieving innovation-driven development; define the functions and roles of enterprises, research institutes, universities and colleges, social groups and other innovation entities, and set up an open and efficient innovation network; improve innovation governance, further clarify the respective responsibilities of government and market, and build a system for ensuring balanced allocation of innovation resources; formulate pro-innovation policies, improve the legal system to protect innovation, and cultivate an innovation-friendly social environment for boosted creativity of the whole society.

Six transformations means the transformation in the model of development from an inefficient model focusing on the scale expansion to a more sustainable one focusing on quality and efficiency; the transformation in the driving force of development from traditional factors to innovation factors; the transformation in the industrial sector from the lower-to-medium end of the value chain to the medium-to-higher end; the transformation in the status of innovation capability from being "behind, on a par with or ahead of" other countries at the same time but mostly "behind" to being "on a par with" or "ahead of" other countries; the transformation in the allocation of resources from focusing on R&D to a balanced distribution along the industrial chain, innovation chain and capital chain; and the transformation in innovation entity from an "elite group" of science and technology professionals to the interaction between the "elite group" and the general public involving innovation and entrepreneurship.

2.1.4 Strategic Tasks

We should closely evolve around the core areas to increase economic competing power, the urgent demand in social development and the major challenges in national security, adopt differential strategies and non-symmetric routes, and enhance the task deployment in main areas and key links.

First, we should push forward innovation in industrial technology system, and create new advantages of development. We should speed up the deep integration of industrialization and informatization, take digitalization, networkization, intelligentization and greenization as the basic technical points to increase industrial competing power, to promote cross-field innovation of emerging technologies in all areas, and build a modern industrial technology system that is rational in structure, advanced and practical, open and compatible, independent and controllable and with international competing power, to support and lead the development of emerging industrial clusters with mass breakthroughs in technologies, and promote the quality upgrading in industries.

We should develop a new generation of information network technology, intelligent green manufacturing technology, ecological green, efficient and safe modern agricultural technology, safety, clean and efficient modern energy technology and technologies for efficient utilization of resources and ecological environmental protection, enhance the information base of economic and social development, to push the manufacturing industry to rise to the high end of value chain, and ensure grain security and food security. We should promote energy production and consumption revolution, build a resource-conserving and environment-friendly society and curb the deterioration of the ecological environment, so that people can enjoy a comfortable living environment. We should march into the deep sea and deep space, develop advanced and applicable technologies for ocean and space, and foster the marine economy and space economy.

We should develop smart city and digital social technologies as well as advanced and effective, safe and convenient health technologies to promote people-oriented new urbanization, and address the challenges of major diseases and population aging. We should develop modern service technology that supports the innovation of business model and the subversive technology that leads the industrial transformation, drive the economic form to the high level, and constantly create new industries to create new jobs. Attention should be paid to subversive technologies that may "reset to zero" the existing investment, talent, technology, industry and rules, proactively arrange research and development of cutting-edge technologies in emerging industries, and strive to achieve "overtaking on curved lanes" and "overtaking by changing lanes".

Second, we should strengthen original innovation and enhance supply from sources. The national strategic demand should be combined with goals of scientific exploration, to strengthen the research deployment of global-related scientific issues, enhance the original innovation capability, raise the overall level of scientific discovery, technological invention and product and industry innovation in China, and support industrial transformation and safeguard national security. We should speed up the basic cutting-edge and high technology research oriented to the strategic demand of the state, energetically support basic research for free exploration, and build a batch of infrastructure and large scientific plants and scientific research platforms that support high level innovation.

Third, we should optimize regional innovation layout and build regional economic growth poles. We should focus on the national regional development strategy, promote the rational division of labor in industries with the gathering and flow of innovation elements, and push the overall improvement of regional innovation capacity and competitiveness. We should build the featured regional innovative development pattern, integrate innovation resources across regions, and build demonstration and leading high land for regional innovation.

Fourth, deepen the military and civilian integration, to promote innovative interaction. We should, according to the overall requirements on military and civilian integration development strategy, give play to the important role of scientific and technological innovation in national defense, and speed up the construction and completion of military and civilian integrated innovation system, to form a new pattern of development with military and civilian science and technology deep integration involving all elements and multiple fields and with high efficiency. We should improve the macroscopic coordination mechanism, carry out coordinated military and civilian innovation, advance the integration of the basic elements of military and civilian technologies, and promote two-way transfer and conversion of military and civilian technologies. In military and civilian integration, attention should be paid to technology integration, product integration, capital integration and talent integration, to well carry out the mechanism innovation of "civilian participating in military" and the open innovation of "military converted to civilian".

Fifth, strengthen the innovation main players to lead the innovative development. We should make clear the functions of all innovation players in different links of innovation, stimulate the vigor of the players, systematically raise their innovation capability and consolidate the foundation for innovative development. We should foster the world first-rate innovative enterprises, build world first-rate universities and disciplines and world first-rate scientific research institutions, establish and complete market-oriented new type research and development institutions and built professional service systems for technology transfer.

Sixth, implement major scientific and technological projects, to realize leap forward in main areas. A batch of major scientific and technological projects will be deployed in key areas related to the national security and long-term development.

Seventh, build a high level talent team and consolidate the root of innovation. We should accelerate the construction of teams of scientific and technological innovation leading talents and high-skilled personnel. We should give play to the important role of entrepreneurs in innovation, energetically advocate entrepreneurial spirit, set up social orientation that innovation is glorious and can make fortune, train a large number of innovative entrepreneurs bold in innovation and adventure, and construct the team of specialized, marketized and internationalized professional managers. We should promote education innovation, reform the mode of training talents, and ensure that the fostering scientific spirit, innovative thinking, creativity and social responsibility is carried out in the whole course of education.

Eighth, promote innovation and entrepreneurship, to stimulate the creative vigor of the whole society. We will build and improve the carrier of innovation and entrepreneurship, develop a maker economy, and form a vivid situation of mass entrepreneurship and innovation.

2.2 Core Force in the Leap-Forward Development of CHSR

The importance of HSR to China's rise in the 21st century cannot be overestimated. HSR is a big industry that can change the world political and economic pattern since China's reform and opening up, is a typical representative of the "wholly" localization of the modern equipment industry of China, and also a main symbol of China marching towards a global power.

Compared with countries that took the lead in developing HSR, the development of HSR in China has experienced a process of "lagging too far behind to catch up with", "following the others", then "running side by side" and finally "taking the lead", realizing the leap forward from "technology import" to "made in China" and then "created in China".[3] In the course of "import" to "taking the lead", China worked continually in digestion, absorption and re-innovation, and now it comes to the "4.0 era" to define HSR by itself.[4] Guided by the innovation strategy of the state, the following five forces have played critical roles in the leap-forward development of CHSR.

[3]Fei Xu: *Going Global Strategy of China High-Speed Railway: Purpose, Strategy and Actions, China Engineering Sciences* No. 4, 2015, pp. 4–8.

[4]Xie Xiaowen: *China should be Confident in Defining High-Speed Railway 4.0 Era, China Economic Weekly* No. 42, 2015, pp. 74–76.

2.2.1 The Institutional and Administrative Forces Dominated by the Government

As a later-comer in HSR development, the Chinese Government gave full play to the institutional and administrative forces, to dominate the leap-forward development of HSR in China in both strategy and policy.

Top level overall planning and arrangement in advance, to define the technical line of indigenous innovation. The Chinese Government did top level design in good time and effective overall industrial planning on the basis of the conditions of the country and its railway. In 2004, the State Council approved and issued the *Medium and Long-term Railway Network Planning* and the *Minutes of Meeting on Relevant Issues in Research of Railway Locomotive and Vehicle Outfitting*, plotting the development blueprint for CHSR with "four longitudinal and four transversal lines" at the national strategic level, and set the technical line of "import, digestion, absorption and re-innovation" for the locomotive and vehicle equipment for CHSR. In 2008, the former Ministry of Railways and the Ministry of Science and Technology signed the *United Action Plan on Indigenous Innovation for High-Speed Trains in China*, further pushed forward institutionally the rapid change from following and imitation to indigenous innovation for CHSR.

Unify the market and focus on import, to start again from the shoulders of a giant. The competent department of railway integrated the domestic market with strong power, and started negotiations with foreign companies with unified principles. By designating strategic buyers and setting the import principles that "critical technologies must be transferred, the price must be the lowest, and Chinese brands must be used", not only avoiding the possible awkward situation that the domestic enterprises are conquered one by one in the process of industrial catch-up, but also breaking the technical "moat" intentionally held by the powerful HSR countries with advantageous position and at low cost, promoting the domestic companies to dock with the international most advanced technology platform at high level, and laying the solid foundation for indigenous innovation by CHSR at a high start point.

Build platforms and direct cooperation, for collaborative innovation by the government, industry, education and research sectors. Under the condition that the market economy in China is to be further completed, innovation elements cannot be effectively integrated without the guidance by the "tangible hand" of the government. The railway competent department arranged main points of research and development with overall planning, defined the key areas of HSR technologies, built the rapid ad efficient innovation platforms for combination of production, education and research, and guided the enterprises to carry out cooperation with universities and colleges and research institutions on specific projects. In the meantime, scientific and technological input for the basic and application researches in HSR was increased with the support by the "Science and Technology Support Program", "863 Program" and "973 Program".

2.2.2 *Market and Industrial Forces of Enterprises as Main Bodies*

Enterprises are the main bodies of market and industry, and more the main bodies of technological innovation. The HSR enterprises of China made full use of their main body position, and of the decisive force of market that allocates the innovation resources, pushed the development of CHSR by closely combining with the industrial force, thereby realizing the shift from "blood transfusion" to "blood making", and from exogenous technical accumulation to endogenous capability.

Guide the innovation direction of HSR enterprises with the force of market demand. The complicated geological and climate conditions of China gave rise to the demand of HSR for multiple type technical adaptive reform, the operation technology demand for long distance, high density and different speeds for operation on the same line or across lines, the diversified testing demand in the course of digesting and absorbing the imported technologies, and the multi-level integrated technology R&D demand resulted from integration of highly comprehensive systems. All these demands pushed from the demand side the HSR enterprises to carry out targeted R&D and innovation of HSR technologies with actual questions as the guide and meeting the market demand as the objectives.

Increase the indigenous innovation capability of HSR enterprises with the market competition force. Orderly competition in the industry was carried out on the basis of market principles with the former CSR and CNR and three main technical platforms as the representative, to promote the enterprises to realize digestion and absorption of imported technologies in an inverted pattern on the basis of relying on production, teaching and research cooperation, and also promote the indigenous innovation in HSR with forward design. In this process, the development, testing and manufacturing capability of HSR enterprises was raised in all aspects, and the capability, efficiency and strength of indigenous innovation fully enhanced.

Expand the innovation space of HSR enterprises with the force of market scale. The large scale HSR construction in China also provided the enterprises with important technological application opportunities and massive "error testing" pilot market for their innovation while bringing them with huge market demand and steady market expectation. The market space and possibility of enterprise innovation was greatly increase with the accumulation of innovation experience in the cyclic process of "identifying problem—solving problem" in the large scale practical market operation and technological applications.

Lever the innovative development of the whole industrial chain with the force of industrial collaboration. HSR, as a knowledge and technology intensive industry, its innovation is interdisciplinary, cross-specialty and cross-industry complex system engineering, and it needs the coordinated development of related industries and cooperation with each other on the whole industrial chain of survey and design, project implementation, equipment supply, completion inspection and acceptance, joint testing and commissioning, safety evaluation, operation management and so on. We should deploy the innovation chain with the industrial collaboration force along the industrial chain, to carry out technological import, digestion and absorption and localization on the whole industrial chain, and then break the difficult points of key technologies and main associated technologies, to promote the innovative development and overall upgrading of the whole industrial chain of HSR and upgrade the overall level and subsequent development capability of key technologies.

2.2.3 The Scientific and Technological and Talent Forces of Integrated Learning, Research and Production

It is the key to successful innovation of HSR to give play to the comparative advantages of universities and colleges, scientific and technological institutions and enterprises (industries) on the basic research, application research and development research, bring into full play the main forces of science and technology and talents, and build the open innovation platform combining learning, research and production with market demand as orientation and the scientific research projects as backup.

Universities and colleges, as important institutions to produce knowledge and foster talents, have played an irreplaceable role in the scientific and technological innovation of CHSR. Especially, research-based universities with features of rail transportation sector have not only fostered railway professional personnel of various levels, but also provided specific scientific and technological support to the sustained development of HSR with their special advantages accumulated in their historical development on aspects of advanced concept as the leader, the principle and basic theoretical research and construction of testing platforms.

Scientific research institutions, as the main undertakers of application-oriented basic research, have given powerful support to the construction of technological innovation capability of HSR enterprises in China on the research of application technologies, simulation testing and verification and improvement of manufacturing techniques with their good scientific and technological facilities, scientific research environment, scientific research talents and research foundation.

The HSR enterprises played the core role as main body in the enterprise, school and research cooperation, brought into full play their respective comparative advantages on product development, market opening and engineering and industrialization with actual innovation demand as the orientation, and cooperated effectively with universities and colleges and research institutions in the forms of talent exchange and inquiry, joint application for subjects, joint construction of laboratories and research platforms, and entrusted research and development.

For the independent scientific and technological innovation in CHSR, the universities and colleges, research institutions and enterprises were well combined, to link science and technology with economy, innovation results with industries, and innovation projects with real productive force, thus accelerating the industrialization of scientific research innovation results. More important, the scientific and technological innovation pattern of school, research and enterprise integration in CHSR has enabled the three to not only make efforts to break major key technologies and common technologies for the present to support the rapid development of the HSR industry, but also deploy in advance cutting-edge technologies and basic research for the future, to lead the sustained and coordinated development of the HSR industry in the future, as powerful support to the innovation in CHSR.

2.2.4 The Mental Force of Dedication, Opening up and Innovation and the Cultural Force of Learning Widely from Others' Strong Points

Spirit is the source of power and culture is the forerunner of action. For such great achievements scored by CHSR in a short period of a dozen years, the mental and cultural factors contained in it cannot be neglected.

Carry forward the mental strength of patriotic devotion, dare to pioneer, and pursue for excellence of the railway people. For a long time, the railway people of China have shaped their outstanding professional quality, dedication consciousness and patriotic feelings, the "locomotive" spirit and "Qinghai–Tibet Railway" spirit, combined with the outstanding quality of taking the state interest as the most important, bold to take responsibilities, diligent in learning and practice work and strive for innovation, has concentrated into the powerful mental strength of patriotic devotion, dare to pioneer, and pursue for excellence of the railway people, and also the precious mental wealth to realize the leap-forward indigenous innovation development of CHSR.

Advocate the cultural power of all-inclusion and learning widely from others' strong points. In the initial period of HSR development in China, by all-inclusion and learning widely from others' strong points in the opening background of economic globalization, China chose the technical line to link with the world most advanced four technological platforms via technology import, and discarded the technical line to obtain key technologies with full proprietary intellectual property rights by full indigenization. By absorbing the advantages of all technological platforms, learning the strong points from others and mutual learning, CHSR not only has good compatibility with the world advanced technologies in main areas of project construction, MTU, train control and traction power supply, more importantly, it realized the late-mover advantages of integration innovation and overtaking on curved lane as a latecomer.

2.2.5 Deep Integration of the Government, Enterprises and Universities as Main Innovation Bodies

The successful innovation in HSR with Chinese characteristics is the outcome of effective use of late-mover advantages on the basis of the industrial foundation of traditional railway of China and the advanced technologies of railway in other countries in the world. In this process, the main innovation bodies of the government, enterprises and universities and colleges, organically integrated the forces of the institutional and administrative systems, market and industry, science and technology and talents with the market demand of HSR construction as the ribbon and through advantage supplement and close cooperation, formed the "triple spiral" relationship structure that the three of them influence and promote each other and rise in spiral, realizing deep integration.

In this relational structure, the government, as the main body of institutional innovation, led the major policies and the critical actions in technological import and development in the indigenous innovation of CHSR; the enterprises, as the main body of technology innovation, made concrete efforts in promoting product development and market development and the technological development on engineering and industrialization; and universities and colleges, as the main body of knowledge innovation, provided the innovative theoretical knowledge and the common, critical and cutting-edge technologies for CHSR. These three main bodies promoted each other, and produced great strength in cooperation and innovation on the basis of deep integration of policy chain, industrial chain and knowledge chain.

2.3 HSR Experience in the Construction of Innovative Nation of China

Conscientious summary of the successful experience of CHSR in leap-forward indigenous innovation is of important reference significance to other industrial sectors, especially the high-end equipment manufacturing industry with huge market size and complicated product systems. It can also deeply inspire us today in deepening the scientific and technological system management reform, energetically implementing the innovation-driven development strategy, and making the country innovation-oriented (Fig. 2.1).[5]

2.3.1 Give Full Play to the Advantages of the Socialist System and Concentrate Forces on Major Tasks

As a capital-intensive, technology-intensive and knowledge-intensive industry, the High-Speed rail technology of China features great complexity and high comprehensiveness, it is very difficult to achieve breakthrough with a single aspect, and it needs the close cooperation of all sectors of the government, enterprise, school and research based on the top-level design.

[5]Fei Xu: *Adhere to the Indigenous Innovation Road with Chinese Characteristics—Successful Practices of China High-Speed Railway, Truth Seeking* No. 4, 2017, pp. 35–37.

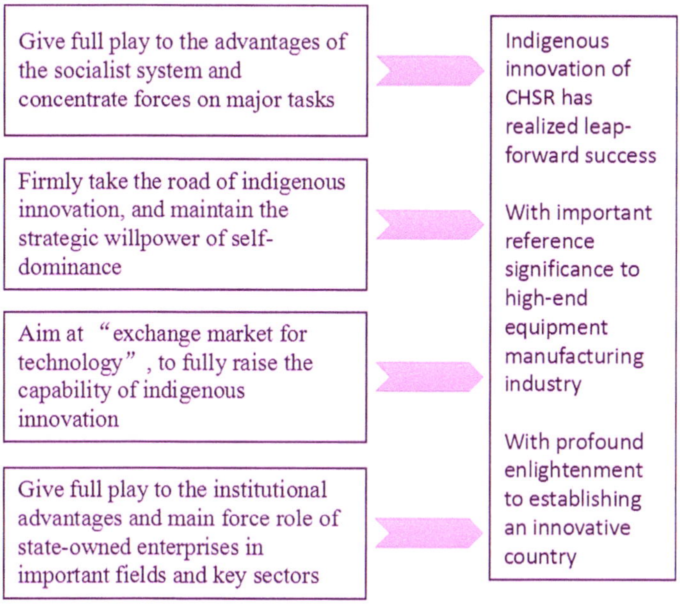

Fig. 2.1 Leap-forward successful experience of CHSR

The *Medium and Long-term Railway Network Planning* approved by the State Council in 2004 is the strategic blueprint for the development of railway transport in China at the national level. Guided by this plan, the competent authorities of railway imported advanced MTU technology from other countries at low cost by institutional means of unified market, unified negotiations with foreign companies and unified product procurement, effectively combined the top-to-bottom top level design with the bottom-to-top subjective initiative by means of scientific research projects, and formed the situation of national unified arrangement to deal with key problems with enterprises as the main body and the enterprise, school and research cross-sector collaborated innovation as the motive force, fully demonstrating the highly effective organization and mobilization capability, authoritative commanding and coordination power and strong action force with unified steps, and fully reflecting the advantage of socialist system that forces can be concentrated to accomplish great tasks.

2.3.2 Firmly Take the Road of Indigenous Innovation, and Maintain the Strategic Willpower of Taking Ourselves as the Main

In the situation that the main nations with HSR in the world remain quite powerful, the important experience that the CHSR could surpass others in a short period of time is the exploration on indigenous innovation with full utilization of global innovation resources as the prerequisite and realizing indigenization of core technologies as the objective. In the era of globalization, the utilization of global innovation resources becomes more convenient, however, core technologies, especially resources that can continually raise the competing power of an industry have become even scarce, difficult to obtain through transaction on open market. Especially, capability cannot be bought, so we must firmly take the road of indigenous innovation, otherwise we would be controlled by others.

The development history of CHSR has shown that, the railway people adhered to the awareness of indigenous innovation in all periods of development, and moulded the real innovation self-confidence with the strategic willpower of taking ourselves as the main. They held this willpower firmly even when they were quite weak while the opponents were quite powerful. China adopted an important policy in importing technologies at that time: all parts must be made in China, even for some parts that must be brought abroad as they could not be made in China for a time being, the foreign company can be included in the list of suppliers only when it has formed joint venture with domestic enterprise. Such policies are vivid interpretation of the awareness and willpower of indigenous innovation.

2.3.3 Aim at "Exchange Market for Technology", to Fully Raise the Capability of Indigenous Innovation

It was not difficult to exchange for the advanced technologies of foreign countries with the huge market size of China, and the difficult points were the digestion and absorption of imported technologies, the breakthrough and mastery of key technologies, and the upgrading of copying imitation to exploration innovation. In other words, we did not simply exchange market for technology, instead, we must master key technologies, and must increase the capability of indigenous innovation and continual innovation, to foster our own brands.

At the very beginning of import, CHSR established the basic policy of "importing advanced technologies, joint design and manufacturing and making the brand of China". Therefore, it not only helped HSR industry of China build a modern manufacturing system, but also obtained the complete experience of product manufacturing and operation, thus enhancing the independent innovation ability of HSR enterprises. On this basis, the tremendous ramifying effect of platform-based technological import and breakthroughs were used, to promote the mass technological accumulation and breakthroughs in the industry chain, thus pushing forward the great development of CHSR in the whole industry chain at the minimum cost and within the shortest period of time.

2.3.4 Give Full Play to the Institutional Advantages and Main Force Role of State-Owned Enterprises in Important Fields and Key Sectors

State-owned enterprises have institutional advantages in strategic and basic technological innovation and in tackling major technical issues. In important fields related to the life line of national economy and key industries with large investment scale, long construction period and slow in obtaining returns, it is particularly important to give full play to the institutional advantages and main force role of state-owned enterprises.

With the national strategy as the orientation, the state will and interests as the bases, the major projects of the state as backup and the powerful organization and calling power as the guarantee, the large state-owned enterprises of the railway industry carried out enterprise, school and research cooperation of advantageous sectors inside and outside the industry, integrated the upstream and downstream of the huge industry chain, and realized the orderly cooperation and win-win benefit under the unified goal; finally, it promoted the import, digestion, absorption and re-innovation with advanced technologies from other countries, broke through the major and key technical difficulties restricting the development of the industry, and established Chinese standards through indigenous innovation, realizing the historical leap-forward from catch-up to taking the lead.

2.4 The Global Strategic Value of China High-Speed Railway

Facing the future from our present time, and seeing the CHSR in a global view, much strategic value can be found (Fig. 2.2).[6]

[6]Fei Xu: *Global Strategic Value of China High-Speed Railway, Frontiers of Academia* No. 2, 2016 (No. 90 total), pp. 6–20.

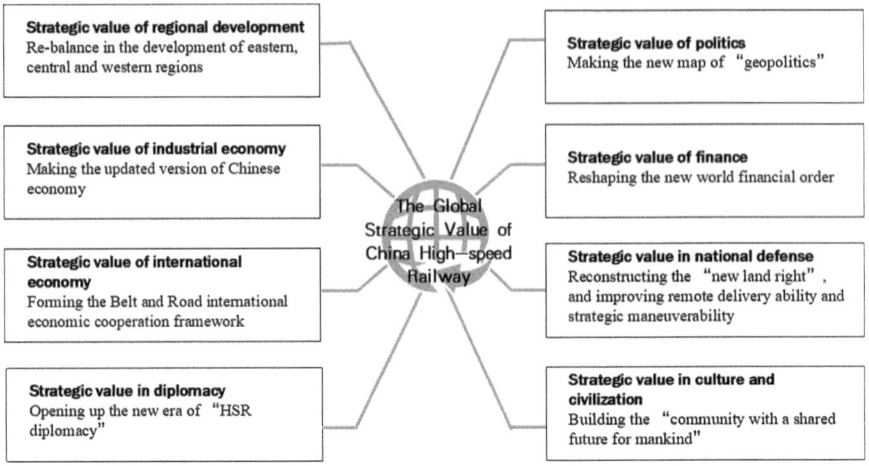

Fig. 2.2 The global strategic value of China High-Speed Railway

2.4.1 Strategic Value of Regional Development: Re-Balance in the Development of the Eastern, Central and Western Regions

It is known to all that regional development imbalance is a major structural problem faced by China's national governance. It is regrettable that the decade of development of the western region has also been a period of rapid expansion of the western and eastern gap. The per capita GDP gap was widened from RMB 7000 yuan in 2000 to RMB 21,000 yuan in 2010. To change the imbalance of regional development, we need to reorganize the whole land and resources in the sense of "the whole nation as a board of chess". Here, we first need to break through the bottleneck of communications and transportation, to transcend the spatial differentiation effect caused by geographical separation.

With the development of China's High-Speed Rail network with "four longitudinal and four transversal lines", rapid flow and distribution has been realized for the streams of people, logistics, energy, cash flow and information flow, thus promoting the optimized allocation and agglomeration development of the capital, technology, labor and other production factors, and consumption elements such as consumer groups and consumption materials, bringing about the vigorous economic and social development of all node cities along the way, and producing important effect on the readjustment of the industrial and economic structures in the eastern, central and western regions.[7]

[7]Xu Yong: *Initial Exploration on Financial Service in High-Speed Railway Era, Financial Times* Jan. 17, 2011.

Today, it takes only more than 4 h to travel from Shanghai to Wuhan with the HSR between these two cities; a HSR train is set out every 5 min from Wuhan to Zhengzhou; and the north-south HSR channel from Hefei to Fuzhou has lined Anhui Province and Jiangxi Province with the Cross-strait Economic Zone.

2.4.2 Strategic Value of Industrial Economy: Making the Updated Version of Chinese Economy

High-Speed railway is an important facility of a big nation. HSR is a large industrial cluster of systemic industry, with a long industrial chain, and it is also a capital intensive and technology intensive industry, therefore it has significant stimulating effect and spillover effect on the industrial economy. The World Bank calculation of drive and spillover effects of HSR with 2010 as the base shows that the contribution of Beijing–Shanghai line to the GDP of Jinan and Dezhou is 0.55% and 1.03% respectively, and the contribution of Changchun–Jilin line to the GDP of Jilin is 0.64%. The HSR increased GDP of Jinan, Dezhou and Jilin by 3.65 billion yuan, 3.59 billion yuan and 2.39 billion yuan respectively.[8]

In fact, the High-Speed rail has the most obvious drive and spillover effect on the infrastructure, manufacturing, and resources development, and the most direct effect on the six industrial sectors of construction contractors, power equipment, engineering machinery, railway equipment, communications equipment and logistics, and the first is infrastructure, including the engineering machinery, cement and building materials involved in the construction of bridges, tunnels and stations. The second is iron and steel, rail production and processing, and machine tool equipment involved in track laying, as well as the procurement of vehicles and associated facilities in the early period of operation, including the production of locomotives and rolling stocks, the electrification information and signal equipment and computer control system. In the final operation and maintenance section, benefits will be obtained continually from the locomotive parts, maintenance supplies and station operation.

The influence of HSR on tertiary industry, especially tourism, is also becoming more pronounced. According to the *Analysis of Regional Economic Effect of China's High-Speed Railway* published by the World Bank, in 2012, the HSR touring groups generated 7–8% of the total revenue of travel agencies, and this figure was about 10% in 2013. The reason is that High-Speed rail can increase travel speed by 50–100%, thus expanding the market coverage of business people. The proportion of business travel among HSR passengers is very high (40% in Changchun–Jilin line and 63% in Beijing–Shanghai line), much higher than that of ordinary railway.[9]

[8] World Bank Report: *Analysis of Economic Impact in High-Speed Railway Areas of China*, quoted from *Only One Day from Hefei to Fuzhou*, *Xinmin Weekly*, October 26, 2015.

[9] *What has High-Speed Railway Brought to us? World Bank Published "Analysis of Economic Impact in High-Speed Railway Areas of China"*, China Transportation Information Network, October 27, 2014.

The technological innovation in the whole industry of HSR will also drive the development of associated industries, build the global HSR production system, mould the world brand of Chinese products, and increase the global competing power of Chinese industries, thus making the "Upgraded version of Chinese economy". CHSR "going global" will promote the export from China to shift to high-end manufacturing with high technical content and intensive capital, and from product trade to technology and service trade, to optimize and upgrade the export trade structure of China. In the meantime, it can also realize export of HSR technology and equipment from China, digest the overcapacity in cyclic industries of iron and steel and cement in China, and achieve trade reciprocity with other countries of the world by bringing into play the respective comparative advantages, to promote the coordinated development of international and regional economy.

After the world economic crisis in 2008, the global economic growth was weak, the recovery was unbalanced, and the risks of economic downturn remained. The United States readjusted its global strategy, implemented the "rebalancing" policy of "return to Asia", led the "Trans-Pacific Partnership Agreement (TPP)" and the "Trans-Atlantic Trade and Investment Partner Agreement" (TTIP), trying to regain the global geopolitical, economic and trade advantages, and offset the development achievements of China's reform and opening up.

The severe international environment forced us to continue to orient to the east, and implement a new round of opening-up to the outside world, and at the same time adjust our strategy, to march to the north, west, south, and deploy for CHSR to "go global", so that the Regional Comprehensive Economic Partnership (RCEP) and the new silk road economic belt can bring out the best in each other. The HSR "going global" is a strategic need for the Chinese economy to get through the close siege, a strategic choice for the Chinese economy to make further progress after more than 30 years of High-Speed growth, and also an important route to realize the "Chinese Dream".

2.4.3 *Strategic Value of International Economy: Shaping the Framework of Belt and Road International Economic Cooperation*

Along the Belt and Road, except for a small number of countries like Singapore, most developing countries need large scale construction of infrastructure as their social and economic development level is still low. China's surplus capital and production capacity, especially the advanced and reliable infrastructure construction technology and equipment, are just needed in these countries. This natural supply and demand relationship is the internal driving force for the economic cooperation in Belt and Road countries. In recent years, CHSR has continued its rapid growth, continuously enhancing its independent development capacity and core competitiveness, and producing a significant influence internationally. Also, the readjustment of global industrial structure and industrial layout have been accelerated, and new industrial chains, value chains and supply chains are emerging, also providing a broad international space for HSR to participate in international production capacity cooperation.

At present, the planning of the six economic corridors and nine overseas channels has built up the physical framework for Belt and Road economic cooperation. The six main economic corridors of China–Mongolia–Russia, new Asia–Europe Continental Bridge, China–Central Asia–West Asia, China–Indo–China Peninsula, China–Pakistan and Bangladesh–China–India–Myanmar,[10] have basically formed the continental framework of the silk road economic belt. Among them, the "China–Pakistan corridor" is focused on petroleum transport, the "Bangladesh–China–India–Myanmar corridor" emphasizes the trading with ASEAN, the "new Asia–Europe Continental Bridge corridor" is the main direct logistics route from China to Europe, and the "China–Mongolia–Russia corridor" stresses the national security and energy development. The nine main communications projects for the Belt and Road planned by the state Ministry of Transport and Transportation, namely China–Laos–Thailand, China–Mongolia, China–Russia, China–Pakistan, China–Kyrgyzstan–Uzbekistan, China–Kazakhstan, China–Tajikistan–Afghanistan–Iran, China–India, and China–Vietnam, have basically formed the silk road channels linking with the main transport lines domestically and extending to various places in the world globally.

The HSR construction for the China–Pakistan economic corridor and Bangladesh–China–India–Myanmar economic corridor will get through the economic channel to the Indian Ocean, directly heading to the world largest petroleum production and supply region—the Persian Gulf. This is the inevitable demand from the overall opening up pattern of China, and also the realistic need in solving the unbalanced domestic economic structure and realizing the overall development of Chinese economy. It is also an economic cardiotonic for the global economy that has been weak, unbalanced in recovery with remaining risk of downturn after the financial crisis in 2008. Therefore, HSR has the mission of rebuilding the Silk Road, with the strategic value of building the Europe–Asia–Africa economic channels and the Belt and Road international economic cooperation framework.

2.4.4 Strategic Value of Diplomacy: Opening up the New Era of "HSR Diplomacy"

Diplomacy needs powerful assistors, and there has been no exception from the past to the present in China and other countries that a backward country has no diplomacy. Since the founding of the People's Republic of China, China's diplomacy has gone through three eras of "ping-pong diplomacy" to "panda diplomacy" and then to "HSR diplomacy". "Ping-pong diplomacy" created a successful model of

---

[10]National Development and Reform Commission, Ministry of Foreign Affairs and Ministry of Commerce: *Vision and Actions to Push Forward the Joint Construction of Silk Road Economic Belt and 21st-Century Maritime Silk Road*, Xinhua News Agency, March 28, 2015.

"small ball driving the earth" and promoting the exchange of nations with people-to-people friendship, enabling China to win equality and dignity on the international stage. Since the reform and opening up, "panda diplomacy" has shaped China's good international image and improved its security environment.

As the wheel of history entered the 21st century, competition among countries in the world has been intensifying, and the strategy of peaceful diplomacy faces more test of strength. In the past "ping-pong diplomacy" and "panda diplomacy", China was in a weak position in diplomacy with unequal strength towards other countries, while in the current "HSR diplomacy", China is at an advantageous position. In recent years, the "CHSR", with major equipment, technology integration, industry association, international trade, international consulting, international contracting, overseas investment and inter-government cooperation and package agreement as new contents, has attracted the wide attention all over the world, and become a new medium for active diplomacy with China at a dominating position. It marks the beginning of the 3.0 era of China's diplomacy, which is commensurate with the status of China as the world's second largest economy.

Some scholars believe that the 21st century can be either a Pacific century or a Eurasian continent century, and even possible a century for both co-existing.[11] It depends on the external environment faced by China and the position and attitude of all action parties towards China. If the Pacific Rim countries are hostile to China, China will go to the west and focus on economic integration of the Eurasian continent; if the Eurasian countries are hostile to China, China will go to the east and engage itself in the integration of the Pacific Rim economy; if both sides show good to China, China can link the Asia–Pacific region in the east and Western Europe in the west, and also go across Africa to form a ring with Asia and Europe, to promote multilateral development. Here, the HSR will become an important diplomatic grab for strategic hedging.

2.4.5 Strategic Value of Politics: Making the New Map of "Geopolitics"

In 1887, British geographer Mackinder presented his article *The Geographical Pivot of History* at the Royal Society of Geography, proposing his famous "heartland theory". His core view was summarized by people later as the famous "those who rule the Eastern Europe can dominate the heartland of the world; those who rule the heartland of the world can dominate the world island; and those who rule the world island can dominate the whole world."[12]

---

[11] Yama: *Great Wisdom of High-Speed Railway and Belt and Road Strategies of China—A Special Interview of Gao Po, Director of China High-Speed Railway Strategy Research Center of Southwest Jiaotong University, Decision and Information* No. 4, 2015, pp. 8–20.

[12] Halford Mackinder: *The Geographical Pivot of History*, the Commercial Press 2015 Version, p. 5.

The deep logic behind the key words "heartland" and "island" in Mackinder's core view is the competition between land right and marine right. The gaming between land right and marine right has always been an important point of concern for big powers. However, the ultimate goal of gaining the marine right is to gain the land right for continent and islands, because humans live on the land after all. The large-scale construction and "going global" of CHSR will support the "land right" with the "right of way", enabling China to enter the "new land right" era, thus inevitably bringing about change in pattern of the marine right and land right in the world, resulting in a huge geopolitical effects.

China has a vast land territory, with a land boundary of more than 22,000 km, bordering with 14 countries on land, and facing 6 countries across ocean. The planned CHSR going global includes three strategic orientations of Eurasia, Central Asia and Transasia. The construction of the Eurasian High-Speed rail will change China's long dependence on marine shipping in foreign trade. The Central Asian High-Speed Railway will run through regions with rich resources, just along the "Silk Road Economic Belt", therefore it will play an important role in expanding the economic and trade cooperation with Europe and inland African countries, speeding up the formation of the golden trunk line of Silk Road Economic Belt, and pushing the "geopolitics" with the "geoeconomy". The construction of the Trans-Asian High-Speed Railway will promote the political stability in the Asia–Pacific Region, and help China to get through the marine outfall in the south, forming the new pattern to open to the Indian Ocean, and further enhancing the international political position of China. From the perspective of geopolitical relations, the CHSR "going global" can also promote the mutual respect and equal consultation politically, mutual trust and strengthened cooperation on security, mutual benefit and complementary advantages economically with neighboring countries, so it is a major initiative of China to participate in reconstruction of the political world order with a positive attitude.

In its reform and opening up, China relied mainly on opening to the east (the Pacific) to join the global economic cycle, and the economic center of gravity of China moved to the coastal region, achieving unprecedented economic development since the founding of New China. However, with the return of the US to the Asia—Pacific Region, China's opening to the east (the Pacific) will possibly be besieged and squeezed. If the three lines of Eurasian High-Speed Railway, Central Asia High-Speed Railway and Trans-Asian High-Speed Railway can be built and put into operation successfully (Fig. 2.3), China will be able to break the containment in the Pacific Ocean by the United States and its Allies in the three directions of the west, the southwest and the north, hedging to a great extent the realistic pressure on China produced by the return of the United States to the Asia–Pacific Region. More than that, China can also undertake resources and technology exchanges with Europe, while making outbound investments in central Asia, western Asia and Africa.

In recent years, the Crimean issue has caused the tension and deterioration of the Russia international relations between the United States, Europe and Russia, resulting in profound changes in the bilateral relations, multilateral relations and international political order between important countries and regions including China, the United States, Europe and Russia as well as the world geopolitical map, while the "Shanghai

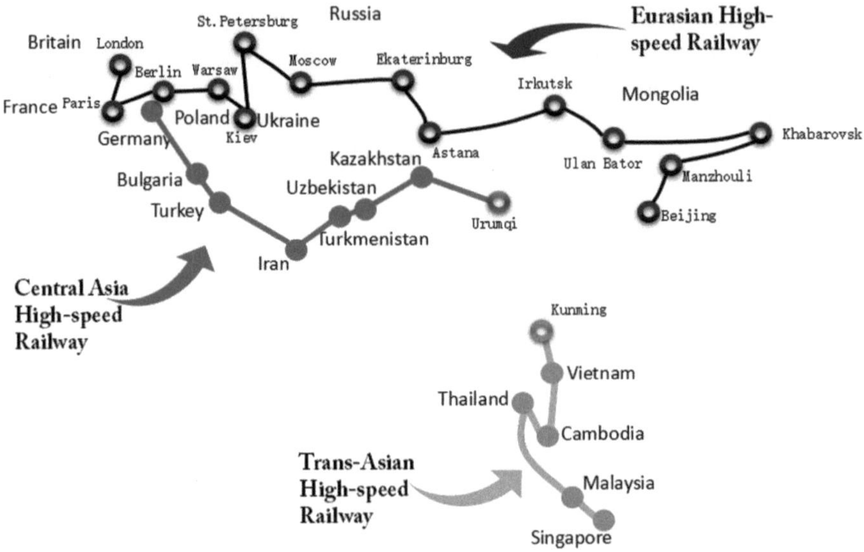

Fig. 2.3 Main countries and cities to be linked by the Eurasian High-Speed railway, Central Asia High-Speed railway and trans-Asian High-Speed railway

Cooperation Organization" with China, Russia and Central Asian countries is also gradually shifting from anti-terrorism and security to economic cooperation and alliance. Reviewing the international economic environment, China cooperated with the SCO member countries and observer countries deeply, on the one hand, China (Shanghai) Free Trade Pilot Zone is used as the forerunner to cope with the TPP and TTIP; on the other hand, the advanced transportation means of High-Speed railway is used to rebuild the "Silk Road economic belt" running across the Eurasian continent.

To sum up, the development quality and volume of CHSR "going global" strategy is a decisive factor whether China can reshape the Eurasian geopolitics new pattern, and will reshape the "geopolitical" new map of the world.

2.4.6 *Strategic Value of Finance: Reshaping the New World Financial Order*

It is well known that both the World Bank and the International Monetary Fund (IMF), or the Asian Development Bank, are dominated by western developed countries. The financial innovation actions for CHSR to "go global" and the Belt and Road, such as setting up the "Silk Road Fund" and "AIIB", will change to a great extent the current world financial pattern dominated by the west.

The Silk Road Fund and AIIB can not only directly support the "going global" of China's high-end equipment represented by CHSR, at the same time, with the "construction + investment and financing (EPC + F)" pattern, it will dissolve the risk of China's huge foreign exchange reserves, speed up the process of RMB internationalization, and provide liquidity for the Asia–Pacific Region, Eurasian and African developing countries in urgent need in their economic development. Deeper level values lie in seizing the right to set and dominate new rules of global trade, and the right of pricing and resources allocation, to lay a solid foundation to achieve regional economic integration on the Eurasian continent and even on Europe, Asia and Africa continent, to become an international economic, financial, trade and shipping center.

On November 8, 2014, President Xi Jinping declared that China will contribute $40 billion to set up the Silk Road Fund, as investment and financing support to projects related to connectivity such as national infrastructure, resource development, industry cooperation and financial cooperation in countries along the Belt and Road. The Silk Road Fund is an important guarantee for CHSR to go global, and also an important trial on the financial support to the construction of the Silk Road economic belt. China's central bank governor Zhou Xiaochuan pointed out that currently the Silk Road Fund has not adopted the concept of sovereign wealth fund, it is an outbound investment fund with foreign exchange as the main, and both foreign and domestic investors can join it by way of marketization. The Silk Road Fund may set up sub-funds for different sectors and regions,[13] to provide financing for Chinese and foreign companies working on the industrial chain for CHSR to go global.[14]

The AIIB, established on December 25, 2015 with its head office in Beijing, is an important strategic deployment for reshaping the new world financial order. Xi Jinping stressed that the main task of the AIIB is to provide fund support to the construction of infrastructure in Asia and the Belt and Road, and is a supplement to the existing international financial system on financing for infrastructure. It is clear that the AIIB is closely related to HSR, because HSR is not only an important infrastructure, but also a forerunner in the implementation of the Belt and Road Initiative.

The initiating member countries of AIIB include not only the developing countries in the Southeast Asia and South Asia, such as Indonesia, Cambodia, India and Bangladesh, but also some developed countries in Europe, such as the UK, France and Germany, and also Australia and New Zealand in Oceania, Brazil in South America, and Egypt and South Africa in Africa. This shows that AIIB is not only an infrastructure investment bank of Asia and a regional multilateral development institution of Asia, in fact, it has become a global financial institution.

The establishment of the Silk Road Fund and AIIB will make up for huge gap existing in the Asian developing countries in the field of infrastructure investment, strengthen the cooperation of countries at different income levels and achieve mutual benefit and win-win progress; It will facilitate the transfer of capital between rich and poor countries and provide appropriate funding for infrastructure construction in

---

[13]Zhou Xiaochuan: *Operations to Launch the Belt and Road Silk Road Fund*, Phoenix Net, February 16, 2015.

[14]Wei Lei: *Silk Road Fund Promotes the Connectivity of Belt and Road, Finance and Accounting for International Commerce* No. 4, 2015, p. 7–11.

poor countries. With the stimulus of the dividend of infrastructure development, it is expected to promote the economic development in Asian and European countries.[15]

Take the Moscow–Kazan High-Speed Railway as an example, with a total investment of 300 billion Rubles (about RMB 32 billion yuan), including 50 billion Rubles as the company's statutory capital investment for the implementation of the project, the remaining 250 billion Rubles will be provided as loans by different banks of China. Another example is the Jakarta–Bandung High-Speed Railway in Indonesia, Indonesia finally chose China instead of Japan, in addition to China's construction speed (the construction period is 3 years, and the railway can be opened to traffic in 2019), what is more important is that Chinese enterprises can provide the loan in full amount, without the need of sovereign credit guarantee by Indonesia.

2.4.7 Strategic Value of National Defense: Reconstructing the "New Land Right", and Improving Remote Delivery Ability and Strategic Maneuverability

At the first meeting of the Central National Security Council, General Secretary Xi Jinping put forward the "overall national security concept" for the first time, and the integration of security on eleven aspects. At present, China's national security in time and space is more extensive than any time in history.

The basic guarantee of national security is the right to control the land, ocean, air, outer space and network. The developed economies in the world are vying to develop aviation and aerospace industries, with the aim at the "right to control the air" and the "right to control the outer space". The United States, with its aircraft carriers as pivot points, has established a huge group of bases around the world, to serves the U.S. global strategy to "control the oceans". In the 21st century, western countries, represented by the United States, have input more forces in cyberspace security. However, the ultimate goal of acquiring the right to control the ocean, air, outer space and syberspace is to guarantee the right to control the land.

In recent years, China has been paying increasing attention to the "marine right", but in the new security situation that the opening to the east (the Pacific) has encountered the containment by the United States, it is necessary to re-examine the relationship of the "marine right" and the "land right", to establish a new land right to hedge the marine right. China should, while firmly implementing the deep blue strategy, take the HSR as the mainstay and support the "land right" with the "right of way", so as to break the siege to advance to the west into the Atlantic Ocean and to the south into the Indian Ocean with the Belt and Road and the six corridors.

This will establish the "land law" to ensure the overall security of the state, to form the hedge of the marine right between the military blockade of the first island

[15]Ding Zhenhui and Huang Xu (City Finance Research Institute of Industrial and Commercial Bank of China): *Asian Infrastructure Investment Bank Opens the Financial Gate for Connectivity, Shanghai Securities News* November 13, 2014, p. A03.

chain and second island chain by the United States while weakening the impact of the United States in the Arabian Sea and getting through the strategic "throat" to the Middle East, and help China get a road of national security with Chinese characteristics. Furthermore, based on the existing and daily enhancing marine right in the Pacific Ocean, we can build up the new marine right based on the Indian Ocean, the Atlantic Ocean and even the Arctic Ocean.

The HSR maintains China's strategic position in the geopolitics of Eurasia and the global landscape, and carries the major missions of international security, economic security, political security and national security. The strategic value of HSR to the national defense can be further discussed from two aspects: strategy and tactics. At the strategic level, the world today has entered the era of "new land right". China is a country by sea, but also a continental country of strategic depth with a vast land, adjacent with many countries on land. China's HSR going global can effectively enhance and safeguard the national security; and the strategic hedging pattern can be established in the game situation of co-existing global maritime right and land right. It can better maintain our position in the geopolitics and the global landscape of Europe and Asia, and better play the important role of a big nation in safeguarding peace in this world that is not quite peaceful.

At the tactical level, long distance delivery capability and strategic maneuverability are important components of the crisis response capacity of a nation and also an important symbol of the comprehensive national strength and military power. China is a country with a vast territory and a long border, including long land boundary and broad coastal border. Therefore it is vital to improve the ability of long distance delivery and strategic maneuverability.

In early May, 2012, Nanjing South Railway Station fulfilled the task to transport light equipment troops as a complete unit for the first time, nearly a thousand officers and soldiers traveled to the training site in south Anhui province via Beijing–Shanghai High-Speed Railway. In late December, 2013, more than a thousand officers and soldiers of the armed police boarded the G216 HSR trains at Shanghai Hongqiao Railway Station. The train ran across 4 provinces, and arrived at the garrison site in North China 1300 km away on schedule.

According to a report by Washington Free Beacon Website on December 21, 2015, the U.S. intelligence services detected the DF-41 missile test of China from a railway wagon. The test was carried out on December 5, a day after the DF-41 carried two sub-missile warheads on a flight test and hit the target the second time on December 4, 2015. The report said that China could use a nationwide rail network to deploy its intercontinental missiles.

In recent years, as more diversified military tasks have been carried out, the demand for long distance delivery and strategic maneuvering of troops has become a normal trend. At present, with the gradual completion of the HSR network in China, the developed HSR network will surely improve the scale and efficiency of delivering mobile combat power various strategic directions in the country, and at the same time, it can also provide great convenience for the strategic maneuverability and concealment for the army.

In addition, the goal of "civil-military integration" is also expected to be realized. As a country with vast territory, and situated in the Asia–Pacific Region with complicated conflicts and interests, it is imperative for China to maintain a certain number of standing troops for a long time for its national security. However, it will also inevitably lead to a huge military spending, as a burden on national economy. It would be quite dangerous for the whole country if the relationship between national defense and economic development cannot be properly arranged. This requires to realize "civil-military integration" as far as possible.

HSR operates mainly on rapid transportation of passengers and electronic commerce at ordinary times, and in time of war, it can transport combating personnel with light equipment, truly achieving "army in the people, civil-military integration, and serviceable at both peacetime and wartime", thus greatly cutting the military spending. With the improvement of the outfitting and informationization level of the army and the breakthrough of transportation and delivery capacity, conditions are ready to reduce the number of military staff.

2.4.8 Strategic Value of Culture and Civilization: Building the "Community with a Shared Future for Mankind"

HSR is a part of transportation and communications. In the Oriental linguistic context, "transportation and communications" can be understood at least at three levels. At the technical level, it is the movement from A to B, namely "transportation". At the cultural level, it is the exchange of ideas, namely "communication". Through communication and dialogue between civilizations, we can promote understanding and consensus, and enhance friendship, so as to realize "harmony under heaven with all people enjoy and share benefits". At the philosophical level, the best interpretation of transportation and communications is the "communication between heaven and earth" in the Book of Changes and Zhuangzi's "All creatures grow with communication". The Chinese philosophers put "transportation and communications" as the way of the universe and the principle of human being.

In short, from "making boat and barge by cutting logs", to the High-Speed rail, "communications and transportation", as people cognized, has realized sublimations from geographical displacement to information interaction, from information interaction to cultural blending, from cultural blending to dialogue among civilizations, from dialogue among civilizations to civilization of dialogue, and finally to the unity of nature and man. In fact, HSR, as a convenient, efficient and green way to travel, has not only greatly compressed the space and time, to make "a bosom friend afar brings a distant land near" a reality, but has also begun and will more deeply promote human communication and exchanges, dialogue and enlightenment between different civilizations, to enhance people-to-people and cultural exchanges of people of all countries along the lines, so that people of different countries can know each other, mutually trust and respect each other, and realize people-to-people connectivity.

No doubt, CHSR going global will open a window to the world to demonstrate the Chinese culture, and show the modern industrial civilization of China on the basis that the world has long known the vast and profound ancient Yellow River civilization and Chinese civilization of China. CHSR going global will also promote the integration of world culture and science and technology. The lines, bridges, stations and trains will all merge the cultural elements of various countries, reflecting the colorful cultural characteristics. From the space structures of the lines and bridges, to the station functions and design concept, and even the car floors and seat fabric designs, new concepts and elements of culture will be contained.

High-Speed rail is an important ribbon connecting countries along the lines, and it will surely promote prosperity and development of countries along the lines. At the same time, with the HSR going global, the Silk Road spirit of "peaceful cooperation, openness, inclusiveness, mutual learning, mutual benefit and win-win" can be passed down from generation to generation. The trans-national development of HSR has weakened the geographic boundary markers of the regional culture in the people's feeling, as the HSR running throughout Asia, Europe and Africa continents, the active East Asia economic circle, developed European economic circle, and the vast hinterland countries with huge potential of development will be more closely linked.

The regional cultures of different countries that were relatively enclosed and lacking in communication in the past have derived new connotation through the complementary integration via High-Speed railway. The High-Speed railway has not only greatly expanded the people's travel and communication radius, but also greatly promoted the understanding, integration and creation of cultures in different regions, and pushed the great prosperity and development of world culture. Although the Belt and Road along which the HSR runs covers a variety of religions such as Buddhism, Taoism, Islam, Orthodox and Christianity in religious belief, and covers the Yellow River civilization, the Ganges River civilization, the Nile civilization, two river civilization and the Mediterranean civilization and other civilizations in forms of civilization, in the long history of these cultural systems have constantly interacted, and their core values tend to be identical.

The routine operation and continuous extension of HSR, on the basis of promoting the core value identity in countries along the way, will fully respect the differences of national cultures of all nations, advocate tolerance of civilizations, strengthen the dialogue between different civilizations, seek common ground while putting aside differences as eclecticism, and form the "friend circle" and "culture circle" that respect each other, blend with each other and learn from each other, to make unique contribution to the building of community with a shared future for mankind of the 21st century.

Part II
The Belt and Road and CHSR "Going Global"

Zheng He, the famous navigator of the Ming dynasty, led a fleet of more than 27,000 people to complete the famous "voyage to the West" in the Chinese history. Today, more than 600 years later, the dream of Zheng He to link with the rest of the world is continuing, and the contemporary development of Belt and Road is promoting the implementation of the national strategy of CHSR going global.

Chapter 3
Belt and Road: Culture and Communications Aiming at Great Unity

In September and October 2013, Chinese President Xi Jinping put forward the initiative to jointly build the Silk Road Economic Belt and the 21st-Century Maritime Silk Road, receiving positive responses from countries along the belt and road. The Belt and Road is originated in China, and links the continents of Asia, Europe and Africa including the Central Asia, Southeast Asia, South Asia, West Asia, Europe and part of Africa, with one side being the active East Asian economic circle, the other side the developed European economic circle, and a hinterland with huge potential for economic development between them, covering 65 countries (see Table 3.1) (of course, the Belt and Road is open, not limited to these 65 countries) with nearly 4.6 billion people, and economic aggregate of about $21 trillion USD, accounting for 64 and 30% respectively of the global total.

The Belt and Road Initiative complied with the common call for cooperation on Eurasia continent, marking a rapid shift of China from an international system participator to a "public products" provider. The "public products" here include not only material public goods, but also conceptual and institutional public products. In recent years, China is committed to promoting the national communication and pragmatic cooperation with countries along the lines, as of March 2017, more than 100 countries and international organizations have responded actively to support it, and more than 40 countries and international organizations signed cooperation agreements with China to jointly build the Belt and Road. Now, the Belt and Road is not only a ribbon connecting Asia, Europe and Africa, but also a "GSM" that connects the whole world.

F. Xu, *The Belt and Road*, https://doi.org/10.1007/978-981-13-1105-5_3

Table 3.1 Countries along the Belt and Road

1 country in East Asia	Mongolia
11 countries in Southeast Asia	Indonesia, Malaysia, Singapore, Thailand, Vietnam, Laos, Myanmar, Cambodia, the Philippines, Brunei, and East Timor
7 countries in South Asia	India, Pakistan, Bangladesh, Nepal, Bhutan, Sri Lanka and Maldives
5 countries in Central Asia	Kazakhstan, Uzbekistan, Turkmenistan, Tajikistan and Kyrgyzstan
20 countries in West Asia	Iran, Afghanistan, Iraq, Turkey, Saudi Arabia, Yemen, Oman, the United Arab Emirates, Qatar, Bahrain, Kuwait, Syria, Jordan, Lebanon, Israel, Palestine, Cyprus, Armenia, Georgia, and Azerbaijan
16 countries in Central and Eastern Europe	Poland, Lithuania, Albania, Estonia, Latvia, the Czech Republic, Slovakia, Hungary, Slovenia, Croatia, Bosnia and Herzegovina, Montenegro, Serbia, Romania, Bulgaria, and Macedonia
4 CIS countries	Russia, Belarus, Ukraine, and Moldova
1 country in Africa	Egypt

Note 44 countries in Asia, 16 countries in Central and Eastern Europe, 4 CIS countries, and 1 country in Africa, a total of 65 countries.

3.1 Silk Road

3.1.1 Silk Road in Narrow Sense

The Silk Road in narrow sense normally refers to the land Silk Road. More than 2000 years ago, the hard-working and brave people on the Eurasia Continent explored a number of routes for trade and humanistic exchange linking Asia, Europe and Africa, and they were collectively referred to as "Silk Road" by the later generations. In 1877 the German explorer and geological geographer Ferdinand von Richthofen, in his book *China*, named the "traffic roads in the western regions between China and Central Asia, and between China and India with silk trade as the media from 114 BC to 127 AD" as "Silk Road", because the roads were mainly for transporting Chinese silk.

The Silk Road was the land trade routes linking the east with the west, and mainly consisted of two directions as the "Northwest Silk Road" and "South Silk Road". The Northwest Silk Road was opened by Zhang Qian of China as an envoy to the western regions, starting from Chang'an (now Xi'an), as a land route extending via Gansu and Xinjiang to Central Asia and West Asia, and connecting the Mediterranean countries (Fig. 3.1). The South Silk Road was an ancient international channel running through Sichuan and Yunnan provinces in southwest China, connecting Myanmar and India, and leading to Southeast Asia, West Asia and Europe, with Sichuan as a starting

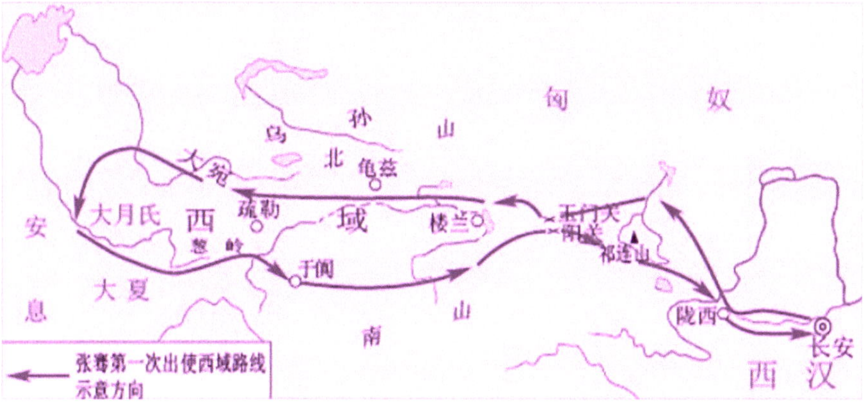

Fig. 3.1 Zhang Qian's trip to the western region. Source: people's education net, http://old.pep.com.cn/czls/js/tbjx/7s/u4/tpsc/201008/t20100820_698286.htm

point, running via Yunnan to Myanmar, India, Pakistan to Central and West Asia, and is a historical route opened after the trip of Zhang Qian to the western regions. After Tang and Song, the South Silk Road gradually became a commercial route typical with tea and horse trade, i.e. the "Ancient Tea Horse Road" extending up to date.

The pioneering work with great hardships on the ancient south and north Silk Road has turned to become the strategic pattern of "Silk Road Economic Belt" today. For thousands of years, the ancient towns, holy places and stations for caravans as stops of merchants, monks and pilgrims were the key links in their vast trade networks. The main corridors of the Silk Road run across inland plateau and mountains in Asia, such as Pamirs Plateau in Tajikistan, Hindu Kush Mountains in north Afghanistan, the Junggar Mountains of Kazakhstan, Kyrgyzstan, Kazakhstan, and Tianshan in Xinjiang of China, and Altai Mountains in Kazakhstan, Russia and Mongolia.

3.1.2 Silk Road in Broad Sense

The Silk Road in broad sense is the collective name of land Silk Road and marine Silk Road (Fig. 3.2). As mentioned above, the land Silk Road was originated with the trip of Zhang Qian sent by the Chinese Emperor Wu of Han to the west and it became the basic trunk road. The "Maritime Silk Road" is the maritime passage of ancient China with foreign countries for trade and cultural exchanges, because it was mainly centered with the South China Sea, so it is also called the Silk Road of the South China Sea. The Maritime Silk Road was formed during the Qin and Han periods, developed in the periods from the Three Kingdoms to the Sui and Tang, became flourishing in the Tang and Song Dynasties, and changed in the Ming and Qing Dynasties, as the most ancient ocean shipping route known to people.

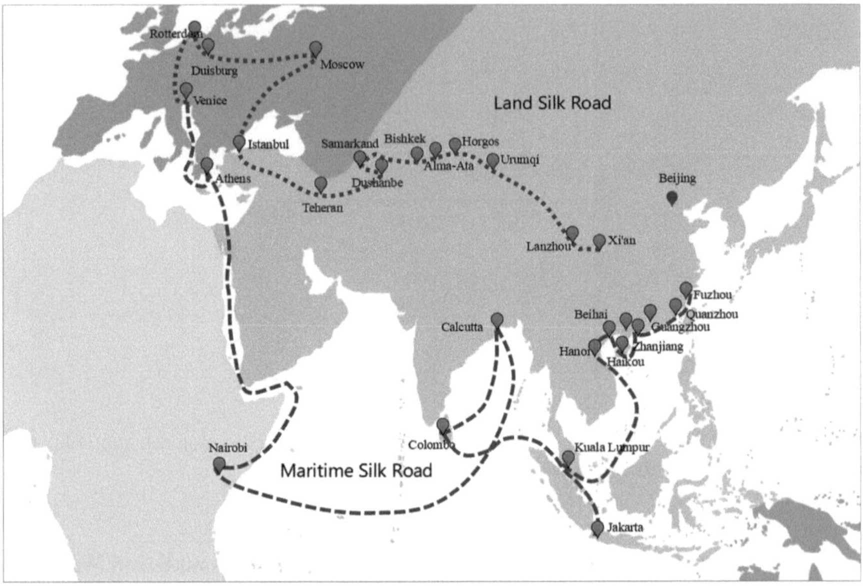

Fig. 3.2 Sketch of Land Silk Road and Maritime Silk Road

On September 7, 2013, Chinese President Xi Jinping first put forward the initiative to jointly build the modern "Silk Road Economic Belt" in Kazakhstan, and on October 3 the same year, he put forward the initiative on the 21st-Century Maritime Silk Road in Indonesia. The Silk Road Economic Belt focuses on unimpeded traffic from China through Central Asia, Russia to Europe (the Baltic Sea), from China via Central Asia, West Asia to the Persian Gulf, and the Mediterranean, and from China to Southeast Asia, South Asia, and the Indian Ocean. The main direction of the 21st-Century Maritime Silk Road is extending from China's coastal ports to Europe via the South China Sea to the Indian Ocean; and from China's coastal ports via the South China Sea to the south Pacific.

3.1.3 The Ancient Silk Road of China

The ancient Silk Road of China was the commercial and trade channels linking with Asia, Africa and Europe, including the official route "Desert Silk Road" of Zhang Qian in the West Han to open up to the West, the "Grassland Silk Road" northwards to the Mongolian Plateau, then extending to Central Asia via the north of Taishan Mountains, the "Southwest Silk Road" from Xi'an to Chengdu and then to India, and the "Maritime Silk Road" starting from coastal cities of Guangzhou, Quanzhou, Hangzhou and Yangzhou, via the South China Sea to Arabian Sea and then to the east coast of Africa. As a traditional land trade power, China created economic prosperity acclaimed by the world via the ancient Silk Road. In the South Song Dynasty, the economic aggregate of China accounted for 75% of the world total, and even in 1820 when China already declined, its economic aggregate still accounted for 33% of the global total.

For more than two thousand years, Zhang Qian went to the west as an envoy twice, Zheng He made seven voyages to the west (Fig. 3.3), Xuan Zang traveled 50,000 li to India and translated 75 volumes of Buddhist classics, and Marco Polo toured in China for 17 years, writing *The Travels of Marco Polo* (Fig. 3.4). Our predecessors in the east and west, riding on camels and horses, and sailing in wooden ships, went through the Taklimakan Desert as the sea of death, crossed the snow-capped Pamirs Plateau and the vast Indian Ocean as geographic boundaries, opened the Silk Road connecting Asia with Europe, established the Maritime Silk Road connecting the Southeast Asia, India subcontinent and the South Pacific, thus opening the ways for economic and trade cooperation, humanistic exchange and civilization integration between the east and west, promoting the material circulation, technical exchange and people-to-people connection and cultural spreading among countries along the routes, drawing closer the geographic, physical and institutional spaces between these countries, and forming a regional connected mass across Asia, Europe and Africa.

In summary, there were mainly four routes of the ancient Silk Road of China: the "Ancient Tea Horse Road" to South Asia via Yunnan and Guizhou, the "Desert Silk Road" via Xinjiang to the Central and West Asia, Persian Gulf and the Mediterranean (Fig. 3.5), the "Grassland Silk Road" via Mongolia and Russia, and the "Maritime

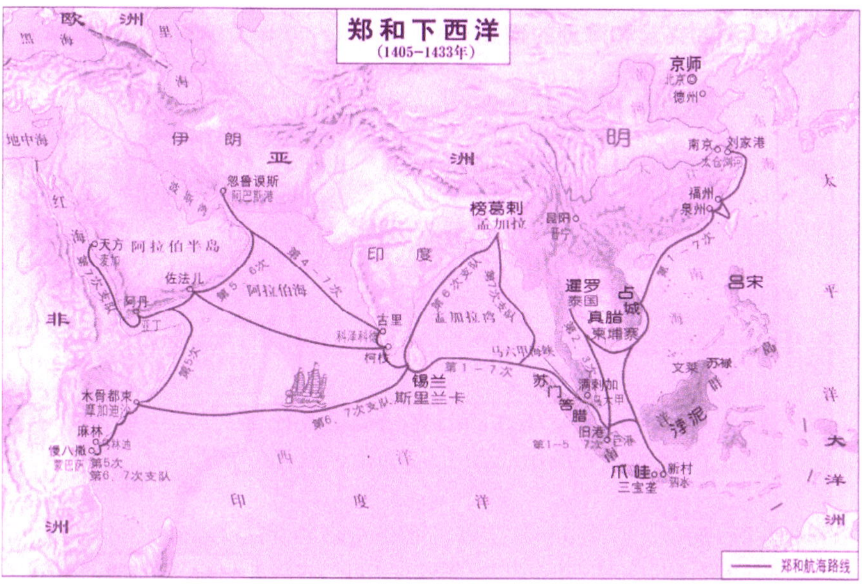

Fig. 3.3 Routes of Zheng He's Voyages to the West (1405–1433). *Source Historical Atlas*, China Map Publishing House and China Map Society, 2016 edition

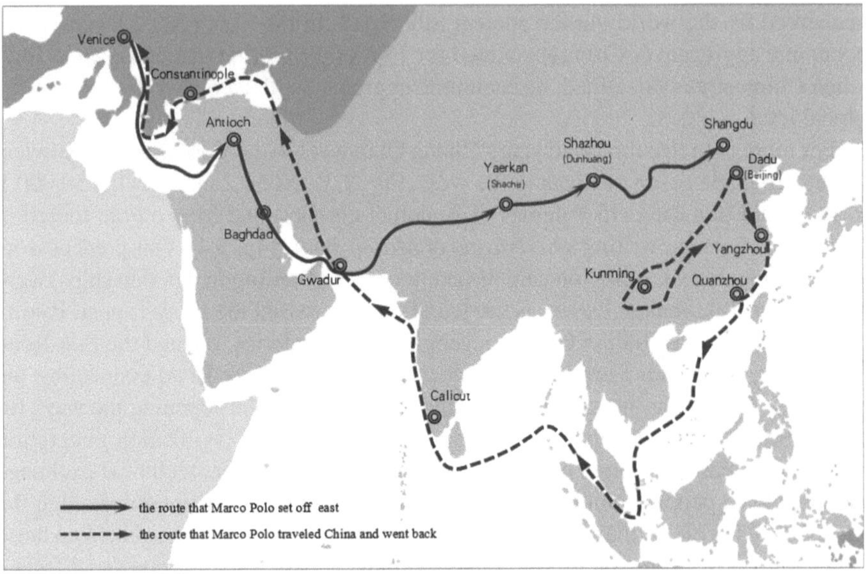

Fig. 3.4 Schematic diagram of Marco Polo's tour lines

Silk Road" via the South China Sea, Pacific and Indian Ocean (Fig. 3.6). "Camel bells in mountains, caravan over the hills, lonely smoke in desert and sunset at a long river", such historical affections, humanistic and meteorological scenes inherent with the nation and evolving with the Silk Road could be hardly felt by countries in other regions. Today, these four channels remain most important ones in China's international cooperation, these channels break through geographic boundaries of Gobi desert, plateaus and snow-capped mountain and vast oceans, and connect Asia, Europe and Africa together economically and culturally through the circulation of commodities, the people-to-people communication and cultural transmission among countries along the lines.

3.2 The Belt Taking the Lead, and the Road Following

3.2.1 Arrangement for Belt and Road: Overall Planning for Land and Ocean, and Starting First on Land

On March 28, 2015, the National Development and Reform Commission, the Ministry of Foreign Affairs, and the Ministry of Commerce of China, with Authorization by the State Council, jointly issued the *Vision And Action to Promote the 21st Century Silk Road Economic Belt and Maritime Silk Road*, which expounded comprehensively the time background, principles, framework ideas, cooperation main

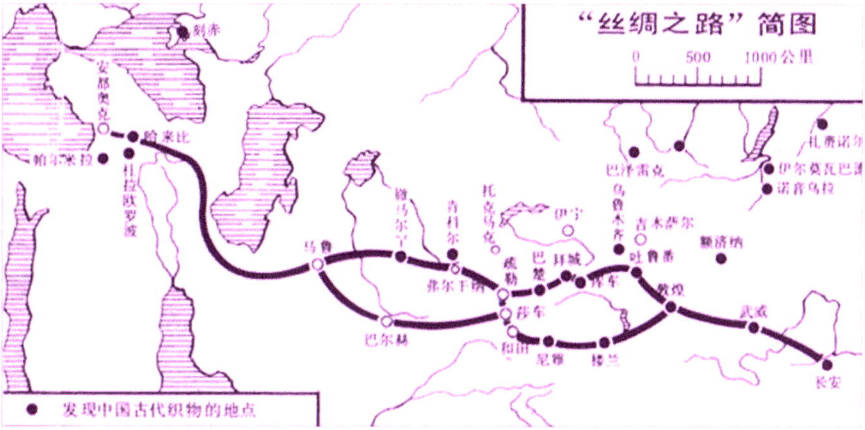

Fig. 3.5 Schematic diagram of ancient Desert Silk Road. *Source* China popular science net, http://www.kepu.net.cn/gb/ezine/ez57.html

points, cooperation mechanism, China's actions and shared vision of the Belt and Road, and made clear that the core in the construction of Belt and Road is to promote policy coordination, facilities connectivity, unimpeded trade, financial integration and people-to-people bonds, build political mutual trust, economic integration, to

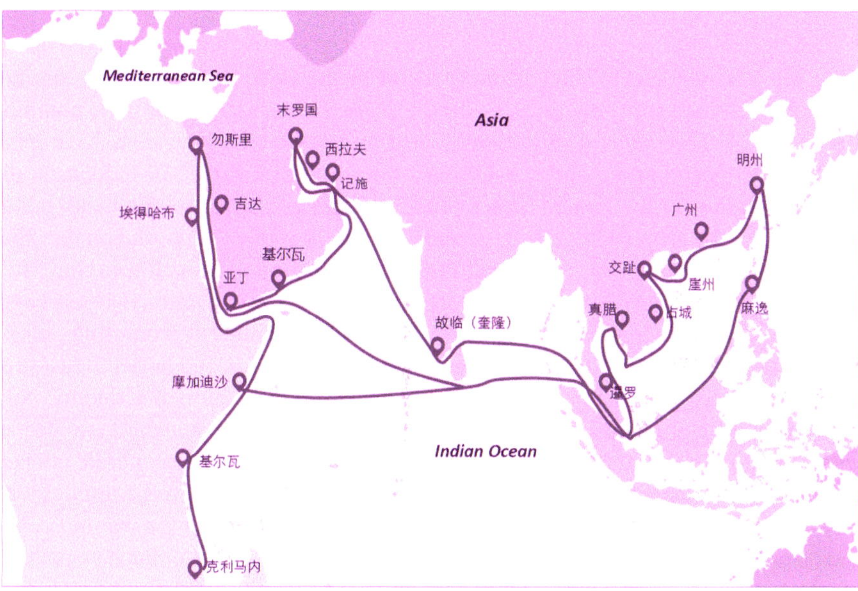

Fig. 3.6 Schematic diagram of ancient Maritime Silk Road

build community of shared interests, shared future and shared responsibilities with mutual political trust, integrated economy and inclusive cultures.

It can be known from the deployment of Belt and Road by China that, "overall planning for land and ocean with action first started on land" should be the pilot deployment with China taking the initiative. The "Silk Road Economic Belt" not only connects the main hinterland of Asia, and its ramifying role and marginal effect will directly act on the smooth progress of the "Maritime Silk Road". The main line of "Silk Road Economic Belt" will get through the big transportation channel from the Pacific to the Baltic, together with the "China–Mongolia–Russia Economic Corridor" in the north, the "China–Pakistan Economic Corridor" in the middle and the "Bangladesh–China–Myanmar–India Economic Corridor" in the south.

In the land and maritime overall planning for Belt and Road, the Belt has the advantages of initiative and selectivity bestowed by nature. Not only the good situation of the AIIB and the steady progress of Silk Road Fund can provide important guarantee on fund for the first actions of the Belt, moreover, the continuous run-in for years by the countries along the belt with the SCO as the platform, the international situation and various factors and relations can also provide important guarantee for the first actions of the Belt.

3.2.2 The "Project Mausam" of India and Opening and Cooperation Development Strategies of Multiple Countries

With the advance and leading demonstration of the Belt, the Road will continually approach the ramifying areas of the Belt, and keep on promoting the common cognitions of the countries in the areas. Just one week after President Xi Jinping put forward the initiative of "21st-Century Maritime Silk Road" in Indonesia, the *Washington Post* in the United States published a long article, on one hand, it was astonished at the proposing of the Belt and Road Initiative by China, and on the other hand, it lamented that the "Silk Road Program" of China overshadowed the "New Silk Road Program" proposed by the Secretary of State Hillary Clinton two years ago.

China's initiative to jointly build the Belt and Road is aimed at promoting orderly and free flow of economic factors, efficient resource allocation and in-depth fusion of market, to push the countries along the line to realize coordination of economic policies, carry out regional cooperation in a wider range, at higher standard and deeper level, and jointly create an open, inclusive and balanced regional economic cooperation framework benefiting to all. Joint construction of Belt and Road complies with the fundamental interests of the international society, and reflects the common ideal and beautiful pursuit of the human society. It is an active exploration on new patterns of international cooperation and global governance, and will add new positive energy for the peaceful development of the world.

Opposite to the active response by the surrounding countries to the Belt and Road Initiative of China, India put forth the "Project Mausam" similar to the "21st-Century Maritime Silk Road", as counter action and strategic hedge. India, as a main big country on the Indian Ocean and an important node on the Maritime Silk Road, is more alert strategically to the construction of the "Road" by China than the construction of the "Belt", and it even believes that the "21st-Century Maritime Silk Road" and the so-called "pearl chain" are the "Silk-Chain" strategy of China to march into the Indian Ocean and then to the vast world.

The "Project Mausam" is designed to compete with the "Maritime Silk Road" program of China by using the advantages of India on history, culture and geography, to ensure the position and role of India in the security and trade in the Indian Ocean region. This project planned a "marine world dominated by India", covering the east Africa, the Arabian Peninsula, extending via south Iran and to the whole South Asia, and to the east, it extends through the Strait of Malacca and Thailand to the whole Southeast Asia.

This region is the influence area of ancient Indian civilization, and the Indian navy will focus on rebuilding such an India-centered order in the region. It should be pointed out that the "Project Mausam" and the "21st-Century Maritime Silk Road" have their respective emphasis, and these two great conceptions have very big space of cooperation and joint construction and sharing. The first realization of the actions for the Belt will directly promote the connection and even integration of the "Maritime Silk Road" with the "Project Mausam", and if well done, the two can run side by side, complement to each other and develop in coordination.

In addition to the "Project Mausam" of India, there are the "Bright Road" of Kazakhstan, "Road of Grassland" of Mongolia, "Europe–Asia Economic Union" of Russia, "Global Ocean Pivot" of Indonesia, "Quadrangle Strategy" of Cambodia, "Turning a Landlocked Country into a Land Connected Country" strategy of the Laos, the "Eurasia Initiative" of Korea, the "Connectivity Overall Plan" of the ASEAN, as well as the "Junker Investment Program" of Europe, the "2030 Vision" of Saudi Arabia, the "Rejuvenation Program" of Egypt, and the four major corridors and trans-border corridor programs and plans of Iran.

In recent years, China has, in the principles of joint consultation, joint construction, joint governance and sharing, connected its Belt and Road Initiative with the implementation of development programs of these countries and regions, and actively promoted the connection of the Belt and Road Initiative with the development programs of the Czech Republic, Poland, Uzbekistan, Brunei and the Eurasian economic union program. In fact, the Belt and Road Initiative not only can closely connect and couple with the development initiatives of different countries, but also highly agree with a number of top level designs such as the UN 2030 Sustainable Development Agenda. The non-exclusiveness, positive externality and the mutual benefit and win-win concept of the communities of shared interests, shared value and shared future of the Belt and Road Initiative have the flexible, sharing and opening elastic space, so it is expected that it will lead the process of new type globalization, and play an important role as "stabilizer" and "propeller" in the world and regional economic recovery and opening.

In the meantime, China has enhanced the function of multilateral cooperation mechanism, and gives play to the role of existing multilateral cooperation mechanism such as Shanghai Cooperation Organization (SCO), China-ASEAN "10 + 1", Asia–Pacific Economic Cooperation Organization (APEC), Asia–Europe Meeting (ASEM), Asia Cooperation Dialogue (ACD), the Conference on Interaction and Confidence-Building Measures in Asia (CICA), China–Arab Cooperation Forum, China-GCC Strategic Dialogue, Great Mekong River Subregion (GMS) economic cooperation and Central Asia Regional Economic Cooperation (CAREC), to strengthen communication with relevant countries and enable more countries and regions to participate in the Belt and Road construction.

3.2.3 *Build the "Asian, European and African" Regional Economies on Both Land and Sea*

The construction of "Maritime Silk Road" is closely related to the various regional and subregional cooperation programs in ASEAN. Especially, the BIMP-EAGA (East ASEAN Economic Growth Area), AMPC (The ASEAN Master Plan for Connectivity) and ARN (ASEAN RORO Network) dominated by the ASEAN and the RCEP (Regional Comprehensive Economic Partnership) dominated by the ASEAN, China, Japan, Korea, India, Australia and New Zealand have countless ties with the construction of the "Road".

As one of the achievements of the Belt and Road Forum for International Cooperation in 2017, the Chinese Government formulated and issued the *Vision for Maritime Cooperation on the Belt and Road Initiative*. According to this *Vision*, mainly three blue economic channels will be constructed: with the coastal economic belt of China as support, connect the China–Indo–China Peninsula economic corridor, and further to the west via the South China Sea into the Indian Ocean, link the China–Pakistan and Bangladesh–China–India–Myanmar economic corridor, to jointly construct the blue economic channel of China–Indian Ocean–Africa–Mediterranean; to enter the Pacific Ocean southward via the South China Sea, to jointly construct the blue economic channel of China–Oceania–South Pacific; and make active efforts to jointly construct the blue economic channel to connect Europe via the Arctic Ocean.

The "Maritime Silk Road" will connect the continents of Asia, Europe and Africa via sea, and finally form a closed loop of sea and land with the "Silk Road Economic Belt" and the regional economic entity of Europe, Asia and Africa, so that the Belt and Road will become the trade channel and economic corridor of Asia, Europe and Africa, ramifying to and influencing the hinterland on both sides and the surrounding areas. The Belt and Road Initiative will achieve connectivity of the Asia, Europe and Africa continents and the oceans by them, establish and strengthen the interconnected partnership of countries along the lines, build up the composite connectivity network in all direction and at multiple levels, to realize the diversified, autonomous, balanced and sustainable development of these countries.

3.3 Integration and Learning from Each Other for Inclusive Development—Extension of the Spirit of Ancient Silk Road in Modern Times

3.3.1 The Spirit of "Silk Road"

The friendly exchanges with countries along the lines are of long standing, starting from "Xu Fu's east crossing" in the Qing Dynasty and the seven sea tours by the Emperor Wu of the Western Han. After "Zhang Qian traveling to the western regions" in the Western Han, business transactions between China and Central Asian and Europe increased rapidly, via the route opened by Zhang Qian, the silk, silk fabric, satin, tea, and porcelain products of China were transported continuously to Central Asia and Europe, and in the Ming Dynasty, the seven voyages by Zheng He to the west connected the marine routes to India, Southeast Asia, Arabic countries and East Africa.

Both Zheng He's voyages to the west in the Yongle Period of the Ming 600 years ago and the contemporary vision of Belt and Road expressed the strong desires of China to achieve connectivity and inclusive development with the countries along the Belt and Road. Zheng He's voyages to the west not only left his name in history, but also made him the mental leader of the contemporary "21st-Century Maritime Silk Road". The Silk Road spirit of "peace and cooperation, openness and inclusiveness, mutual learning and mutual benefit" is the collective memory and historical wealth of all nations along the lines, and the soft power, smart power and sharp power for the success of the Belt and Road Initiative. The Belt and Road Initiative is the extension of the spirit of the ancient Silk Road in the contemporary time, and also the expansion of the historical and cultural heritages shared by all countries in the world in the new era.

It is based on the profound historical affection and heroic modern ambition that the great vision of Belt and Road Initiative that crosses time and space has come from the depth of history, getting through the ancient and the present, connecting China with the rest of the world, complying with the tide of the times for peace, development, cooperation and win-win, carrying the dreams of all countries along the Silk Road for joint development and prosperity, and vesting the ancient Silk Road with the new connotation of the times. For thousands of years, the Silk Road spirit has been passed from generation to generation, and pushed forward the human civilization and progress. It is an important ribbon to promote the prosperity and development of countries along the lines, a symbol of east–west exchange and cooperation, and the historical and cultural heritage shared by all countries in the world. To carry on the "Silk Road spirit", we must promote mutual learning of civilization, respect the choice of roads, adhere to win-win cooperation, and advocate dialogue and communication.

In March 2017, Chinese Foreign Minister Wang Yi pointed out in his reply to question from journalists during the two national conferences that the Belt and Road Initiative was proposed by China, but the opportunity belongs to the whole world.

China is pursuing opening-up, neither Monroe doctrine nor expansionism. The Belt and Road Initiative is an international public product first proposed by China and benefiting all countries. It is not a solo of China, but a symphony participated by all countries. The "Silk Road" represents a spirit and culture of the ocean accepting all rivers running into it, opening to the outside world, learning from each other, exchange, mutual promotion and common development, and this spirit is necessary for any "silk road program" for sustainable development.

In fact, enlightened by the concept of silk road, various "silk road programs", either on sea or on land, formulated by Japan, Russia, Iran and even the United Nations, all demonstrated their strategic ambitions with the Central Asia as the core in different phases, however, such ambitions or strategies have either been discounted or given up halfway, and even ended up with nothing definite, just like the "New Silk Road Program" proposed by the United States in 2011.

There were various causes to such dilemmas, such as harsh preparation, weak implementation, exhausted strategy or change of situation, but the root cause was lacking in the firm heritage and adherence to the "Silk Road Spirit" at the value and concept level. At present, all countries in the world need to trace to the source and absorb nutrient from the "Silk Road Spirit". Looking forward to the future, with the further progress of the Belt and Road Initiative, the Silk Road Spirit will surely be carried on and brightened.

3.3.2 *Social Development Is Originated from Trade Exchanges*

Back in history, the origin and continuing motive force of the Silk Road was the trade of commodities. Accompanied with the trade was the exchange of civilizations and social progress. Porcelain was an important commodity on the Silk Road, and in the past European continental palaces such as the Versailles Palace and Schönbrunn Palace were displayed with fine Chinese porcelain from the Silk Road trade. From Delft in the Holland and Meissen in Germany to Sevres in France, Chinese porcelain gave birth to European porcelain, and the civilizations on both ends of the Eurasian continent resonated here.

The bone china, created slightly later in England, flew back to China as a completely new variety, in turn promoting the new development of porcelain in China. "You give me Papaya, and I take the precious jade in return", the economic exchange promoted the common progress and development of different civilizations. It spread the resources, produces and technologies of economies along the routes in Asia, Europe and Africa, promoted the circulation of all elements, and finally realized the growth of economic performance and improvement of social welfare.

The progress and dissemination of metallurgical technology, aquaculture technology, textile technology, and transportation tools and cooking methods have promoted the technological progress and economic development of economies along the routes.

On the contrary, isolated civilization could only miss good opportunities for its development. China's ban on marine trade in the Ming dynasty and seclusion in the Qing Dynasty, deliberately blocked the communication and contact with the outside world, stuck to the simple agricultural society as described by "living within hail but never visiting each other", simply missed the good opportunity and the situation of social and economic transformation, and resulted in the humiliation for nearly a century in modern Chinese history. It was during the reform and opening up in the recent three decades and more that China took initiative to merge itself with the world economy and trade circulation system, so that China could catch up with them and return to the rank of major economic powers.

3.3.3 Trade Exchanges Promote Cultural Integration

The new resources, products and technologies obtained through Silk Road trading, once integrated into the own cultural texture, would become the endogenous gene for growth. The imperial tea garden of the Western Han at Mengding Mountain in Ya'an of Sichuan is the first place to plant tea trees in the world. In the mid-Tang, the Buddhist Saicho introduced the tea to Japan. Not long after, tea planting was also successful in India and Sri Lanka. The tea trees and plantation technology from China were integrated with the local water and soil and climate, producing high quality tea leaves.

The tea culture of China was also output with tea leaves, and various countries accepted the tea custom of China and developed their local features. Chinese tea was accepted in Britain and turned into British black tea and the traditional afternoon tea; became the teh tarik known as fly tea in Malaysia; and developed into the famous tea ceremony in Japan. The Chinese tea culture, once confined in one place, has now been disseminated all over the world with refreshing fragrance, has developed with various features, prevailing all over the world.

3.3.4 Cultural Integration Promotes People-to-People Bond

"Friends should not be based only on benefit, power or right, and can last long with heart-to-heart communication." Exchange starts from trade of articles and things, and extends gradually to material life and mental culture. However, to realize integration from exchange, people-to-people bond is first required. The people-to-people bond is the social basis for mutual learning of civilizations and inclusive development.

The Daci Temple in Chengdu is the ordainment place of Xuan Zang, a famous monk of the Tang Dynasty. Xuan Zang went to the west via the Silk Road for Buddhist learning, to look for the original allusion in a long journey of more than fifty thousand li, via Gansu, Xinjiang, Uzbekistan, Afghanistan, Pakistan, Kashmir, and India, and studied Buddhist scriptures at many places including the Nalanda

Monastery in India. Back to China, Xuan Zang and his disciples did the painstaking translation of Buddhist scriptures, to carry forward the Buddhist doctrines, so that the Buddhist wisdom took root in the land of China. Buddhism not only became one of the three important branches of the traditional Chinese culture "Confucianism, Taoism and Buddhism", also contributed to the birth of the Neo-Confucianism in the Northern Song period to integrate Confucianism, Taoism and Buddhism, becoming a new form of Confucianism, and opening up a new period of the progress of Chinese civilization.

At the same time, Xuan Zang wrote the *Great Tang Records on the Western Region* in 12 volumes, to comprehensively introduce the geography, produces, customs, culture and features in 110 countries he had travelled in person and 28 countries heard about, enhancing the public understanding of the western countries at that time, and promoting the people-to-people bond across different civilizations. Xuan Zang, with his personal efforts, carried on Buddhism with a great aspiration, and achieved deep integration of the Buddhist culture with the local Chinese culture, not only giving birth to the Chinese Buddhism, but also contributing to the rehabilitation of Confucianism, enabling it to penetrate into the underlying architecture of the Chinese traditional culture. In the process of people-to-people communication, we need not only spiritual messengers like Xuan Zang, but also the joint efforts of many organizations and countries.

To sum up, social development originates from commercial and trade exchange, such exchanges push forward cultural integration, and the cultural integration promotes people-to-people bond. The important implication to us is that, for the Belt and Road Initiative, we should strive to construct the "commercial and trade circle" and "circle of friends".[1]

3.3.5 Build the "Commercial and Trade Circle" for Belt and Road with Road Connection as the Pilot

Our forefathers, under the condition of quite limited resources hundreds of years ago, broke the barriers of high mountains, deserts and oceans, opened the Asia–Europe east–west commercial and trade channels, realized the stable products, resources and personnel exchanges, and achieved the prosperity of the ancient silk road. One of the important contents of the current Belt and Road Initiative is to promote the infrastructure connectivity and international development in the countries along the belt and road, and jointly build international economic cooperation corridors. As an important part of "transportation and communications connectivity" in the area of "facilities connectivity", HSR is an important support for the Belt and Road Initiative in specific directions and regions.

[1] Fei Xu: *Integration and Learning from Each Other for Inclusive Development, Guangming Daily* November 29, 2015, p. 8..

Road connectivity is both the true contents of the Belt and Road Initiative, and also the best entry point and important driving force for the implementation of the Belt and Road Initiative. It is stated in the *Vision and Action to Promote the 21st Century Silk Road Economic Belt and Maritime Silk Road* that, "infrastructure connectivity is a prior field in the construction of 'Belt and Road'". It is an important route in the construction of Belt and Road to promote free exchange of trade with trunk traffic lines or comprehensive transport channels as the development axis and basic carrier of economic corridors.

The Belt and Road Initiative not only has extended the strong desire of ancient silk road for commercial trading and cultural exchange, but also faces new practical demands and challenges. From the domestic demand of China, the "commercial trade circle" of the Silk Road is an important part of the sustainable development strategy. As China's economic development has entered a new normal, new requirements have been put forward for the transformation of development pattern and readjustment of industrial structure. At present, China faces the situation of both excess capacity and excess capital, and is in urgent need to explore overseas markets. The implementation of overseas projects as the HSR going global will spur the export of design, construction, operation and management personnel, promote the shift of domestic construction ability, provide more jobs, and at the same time can produce massive demand for steel and cement. The export of railway equipment will enable the export of locomotive vehicles, MTU, construction equipment, maintenance equipment, communication signal devices, etc., thus promoting the development of other industries. In external demand, most developing countries along the Belt and Road are at a fairly low level of social and economic development stage, China's surplus capital and production capacity, especially the advanced and reliable infrastructure construction technology equipment, can just satisfy the massive demand for infrastructure construction in these countries. This matching supply and demand relationship is the internal driving force and basis for the forming of "commercial trade circle" of the Belt and Road.

Railway transportation, with its large capacity and High-Speed, is undoubtedly the most realistic, economical and reliable choice to realize the "commercial trade circle" of the Belt and Road. CHSR has formed core technology and mature management systems with proprietary intellectual property rights in the engineering construction, High-Speed train, system integration and operation and management as a whole industrial chain, and can serve as the forerunner in the implementation of the Belt and Road Initiative.

3.3.6 Forming the "Culture Circle" of "Belt and Road" with Mutual Progress as the Purpose

The Belt and Road is a great practice with profound historical significance. It is not only the "economic road" that promotes development, but also a "grand way" at the culture and civilization level. The Belt and Road Initiative is shifting from function orientation to the humanistic orientation, emphasizing the purpose of mutual progress and the inclusiveness, focusing on looking for identity contained in diversity and differences, to form the "culture circle" with the greatest common divisor and realize coexistence and co-prosperity of different civilizations. Forming a "culture circle" does not mean making a small circle, and there is no hierarchical "threshold". Coexistence and co-prosperity does not mean the pursuit for rigid convergence or exclusion, neither assimilation nor conquering, instead, we will build the humanistic pattern and civilization ecology for mutual understanding, respect, appreciation, promotion and benefiting.

The European integration process is a model of regional cooperation and regional integration since the formation of national states. The extent of its development up to date was not only out of the expectation of all countries over the world, but also greatly beyond the imagination of European politicians. European integration developed from the initial "Schuman plan" after WWII, the European Coal and Steel Community, the European Economic Community, to the European Community, and then expanded to today's European Union, covering 28 member states, with 24 official languages, and about 500 million people. It contains the diversified cultures of obvious "continental" characteristic French culture, the "ocean" characteristic British culture, and the "extremely conservative and specially advanced" Deutsche culture diversity, yet the EU, as a whole, has a unified "consciousness of Europe". The combination of "nationalism" and "consciousness of Europe" allowed the idea of European Union to be put into practice. Jacques Delors, former chairman of the Commission of European Community, believed that the EC is "the crystallization of ideas and culture".

Although the cultural differences in countries along the Belt and Road are greater than that between Europe countries, the inclusion of multiple culture and the restraint of nationalism by Europe countries in the integration process of about 100 years are worth learning in our process to build the "culture circle" of Belt and Road. Although the "culture circle" of Belt and Road covers three major religions and five cultural systems in the world, these cultural systems are constantly communicating and converging with each other in long historical exchanges, and their core values tend to be identical. For example, the Confucian culture emphasizes "the Benevolent loves others", the Christian advocates "wholeheartedness", "Loving your neighbors as yourself", the Buddhism advertises "Releasing oneself from purgatory" and "Delivering all living creatures from torment, the Hinduism demands "Ahimsa", and the Islamic *Koran* requires people to "be close to neighbors, both near and afar". The civilizations along the Belt and Road have rich cultural diversity, but there is no essential difference in the fundamental value concept, which is the cultural basis for building the "culture circle" of Belt and Road. In the *Doctrine of the Mean*, it is said that "All things grow without harm to each other, and doctrines are parallel without conflict". The construction of Belt and Road requires both national confidence and cultural consensus; both mutual learning of civilizations and cultural integration, so as to dispel doubts and realize common development of different civilizations.

3.3.7 Developing the "Circle of Friends" of Belt and Road with the Aim of Inclusive Development

President Xi Jinping said when meeting with former British Prime Minister Cameron that, the Belt and Road is open, it originates from but is not restricted to the ancient Silk Road, it connects the Asia–Pacific economic circle in the east and the European economic circle in the west, and is a broad "circle of friends" covering Africa and connecting Asia and Europe, all countries interested in it can be added into this "circle of friends". Now the Belt and Road circle of friends are not only growing with increasing number of friends, but also becoming daily active.

The Belt and Road covers over 60 countries and around 100 ethnic groups. In the form of civilization, it involves Yellow River civilization, the Ganga River civilization, the Nile civilization and the Mediterranean civilization; in religious beliefs, there are various religions, including Buddhism, Taoism, Islam, Orthodox and Christianity; and in terms of the extent of development, it covers developed countries, emerging economies and masses of developing countries. To build such a "circle of friends", inclusive development is the only right way and proper course.

In the history of human civilization, inclusive culture was originated in the Mediterranean countries in Europe. The western society with Europe as the center, in its international relations, has experienced three different cultural forms, i.e., the Hobbesian culture characterized by the theory of "original evil" and not tolerant with the "heterogeneous", the Lockean culture with "survival and allowing to survive" as the moral foundation, and the Kantian culture with "All for one and one for all" as the code of ethics. Killing and being killed was the characteristics of the state relations under the Hobbesian culture, competition and cooperation constitute the international relation criteria of the Lockean culture, and rule of law is the normative foundation of the Kantian culture. It is based on the general acceptance by all countries in Europe of the Lockean culture and Kantian culture that Europe became the first region starting the integration process.

In the traditional Chinese culture, "inclusive for dukes" and "gentle to those afar" of the Confucianism, "no rivalry" of the Taoism, "universal love" and "no attack" of Mohism, and "prudent of war" and "thoroughfare and land joining" of the military strategists, are all combined into the political ideal "Great Unity under Heaven". More than 400 years ago, the Italian missionary Matteo Ricci, exclaimed after touring and living in China for over 30 years, "China is not just a kingdom, China is really a world". Actually, this was not the praising words of Matteo Ricci for China. The traditional culture of China has no concept of state in the modern sense of geography. "China" in a cultural concept is "the kingdom of heaven", and it has the inherent inclusiveness.

The ethnic relations and foreign relations formed under the influence of this concept reflect the rational light of "seeking harmony for peace". Therefore, the Belt and Road advocated by China is by no means a "tributary system" of the new era, instead, it is a "circle of friends" formed on the basis of fully respecting the differences of all countries and all nations in the world, for integration with each other and learning from each other. The diversity of civilizations is a basic feature of human society and also an important driving force for the progress of human civilization. It is the essence of inclusive development to respect the various civilizations of different types, regions and stages of development so that they can enjoy equal, symbiotic and free development opportunities.

3.4 The Silk Road Calls for Development of Rail Transport

3.4.1 Rail Transport: The Forerunner for Belt and Road

At the APEC Summit in Beijing, President Xi Jinping said, "if we compared the Belt and Road as the two wings for Asia to take off, the connectivity is the blood veins and channels in the two wings".

In the integrated transportation infrastructure interconnectivity of railway, highway, aviation, pipeline and marine shipping, railway, with its attributes as basic facilities, economy, public welfare, sociality, supportiveness and guidingness, as well as its advantages of safety, convenience, large capacity, and all-weather operation, will surely become the primary choice of transportation infrastructure interconnectivity and priority areas, as the forerunner to boost the construction of the Belt and Road.

There are massive communications demand along the Belt and its nearby areas, the ports of the Road and the key junctures and nodes of hinterland, and the light rails, metros and inter-city railways linking the cities in hinterland. The three main economic corridors for the Belt, i.e. the "China–Mongolia–Russia Economic Corridor" in the north, the "China–Pakistan Economic Corridor" in the middle and the "Bangladesh–China–Myanmar–India Economic Corridor" in the south, will first attract large number of population, driving up the construction of cities and industrial parks.

The demand for inter-city railway construction between node cities in the same economic corridor and between hinterland cities and ramifying cities and for urban mass transit will gradually increase with the progress of the Belt and Road. It should be particularly pointed out that, the "Bangladesh–China–Myanmar–India Economic Corridor" in the south line of the Belt and the economic cooperation of China-ASEAN countries on land, such as the Greater Mekong River Subregion (GMS) economic cooperation, will promote the first connection of the Belt with the Road.

3.4.2 Trans-Asian Railway: The Basic Framework of Belt and Road

As the emerging economic belt with the biggest span and largest coverage in the world, the combination of the Belt and Road Initiative with the regional development and opening-up will completely change the previous development pattern with separate dots and blocks. Transversally, it will run through the east, central and west parts of China; and longitudinally, it will link the main coastal port cities, and extend continually to the Central Asia and ASEAN. We can say that the construction of transportation infrastructure will directly determine the strategic progress speed of the Belt and Road, and railway construction will play a decisive central role.

For the connectivity of Europe and Asia continents, the Trans-Asian Railway (TAR) can become the basic framework of the Belt and Road. The TAR is a unified freight transport railway network connecting the Eurasia Continent. Delegates of 18 Asian countries officially signed the *Inter-government Agreement on Asian Railway Network* in Pusan of Korea on November 10, 2006. According to the planning in the agreement, the gold corridor formed by the four "steel silk roads" in the future will connect Europe and Asia, and the crisscross trunk and branch lines will form a huge network for economic cooperation.

For the TAR, four lines respectively in the north, south, south-north and ASEAN have been determined. Among them, the north and south lines will run through China. The north line is: Korean Peninsula–Russia–China–Mongolia–Kazakhstan, running through the north region of Asia. The south line is: South China–Myanmar–India–Iran–Turkey, connecting the south region. The north–south line is: Russia–Central Asia–the Persian Gulf, connecting the south and north regions. The ASEAN line will link the Southeast Asia region including the Southeast Asian countries and countries in the Indo–China Peninsula. The total length of these four lines is 81,000 km, mostly agreeing with the lines planned for the Belt and Road.

Sometimes, it is customary to call the "Southeast Asia Corridor of the Trans-Asian Railway" also as Trans-Asian Railway for short (Fig. 3.7). It was originated from an initiative put forward by the then Prime Minister Mahathir of Malaysia at the Fifth ASEAN Summit at the end of 1995, namely to build a Trans-Asian Railway beyond the scope of the Mekong River Basin, from Singapore at the southern tip of the Malay Peninsula, via Malaysia and Indo–China to Kunming, China, and this initiative may be originated from the concept of "Trans-European Railway".

3.4.3 Main Channels on Land: Urban Cluster Transportation Demand

On land, the Belt and Road will connect Asia, Europe and Africa as mainly backed by international main channels, to jointly build the international economic cooperation corridors of new Eurasia Continent Bridge, China–Mongolia–Russia, China–Central Asia–West Asia, and China–Indo–China Peninsula with central cities along the lines as support and main economic and trade industrial parks as cooperation platforms. On the sea, major ports will be taken as nodes, it will not only include the inter-continental navigation routes extending to Asia, Europe and Africa, and join with the land route in Italy to form a closed loop of Belt and Road, but also run southward

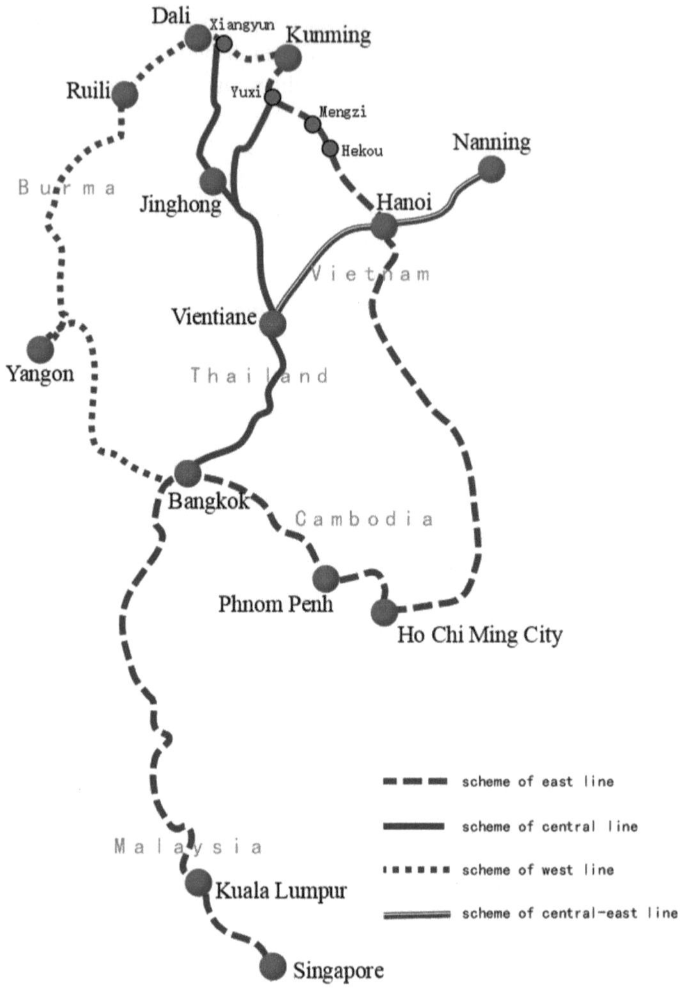

Fig. 3.7 Planning of Southeast Asia corridor of the Trans-Asian Railway

through the densely populated ASEAN, and finally reach the South Pacific region. Along these safe and efficient main transport channels and in the ramifying hinterland areas, city clusters with concentration of population will gradually form, and they will also call for development of mass transit.

3.5 "Big Transportation" in the Framework of Belt and Road

Transportation was the core topic of the ancient Silk Road. In the grand background of contemporary Belt and Road, both the connotation and denotation of transportation have been greatly expanded, and evolving into "Big transportation". The scale of transportation is not only embodied by the movement and the connectivity of land, ocean and air infrastructures at the technical level, but also the integration of infrastructure construction with industrialization and urbanization at the economic and industrial levels, and more the exchange and communication and integration aiming at the formation of the "community with a shared future for mankind" at the culture and civilization levels.

3.5.1 "Big Transportation" at the Physical and Technical Level: Connectivity of Land, Ocean and Air Infrastructures

With the in-depth implementation of the Belt and Road Initiative, breakthroughs have been obtained in the construction of a batch of main channels and critical nodes of transportation infrastructures, and initial results have been obtained. With the construction of the "six corridors and multiple ports in several countries" including the China–Pakistan economic corridor, some important initial achievements were obtained in fields including infrastructure, the connectivity network is being gradually formed, the construction of Hungary–Serbia Railway and Jakarta–Bandung High-Speed Railway has been started, and important progress has been made for the construction of China–Laos and China–Thailand Railways of the Trans-Asian Railway network.

In China-ASEAN direction, two pairs of railway frontiers (Pingxiang–Dong Dang, Hekou–Lao Cai) have been formed, and 8 expressways have been basically completed or planned; in China–Europe direction, there are two main railway lines – the first Eurasia Land Bridge and second Eurasia Land Bridge, extending to Europe respectively via Erenhot, Manzhouli, Alataw Pass and Horgos of China; in China—Central and Western Asia direction, China has built three main land transport channels with Central Asian countries, including 6 cross border highways, 2 railways and 1 pipeline, and opened 8 border ports.

Statistics show that of the countries along the Belt and Road, 9 countries have already been connected with China via railway, 28 countries realized air route connection with China, and 58 countries realized marine route connection with China. They include 4 countries connected with marine, railway and air routes; 21 with marine and air routes, 4 with marine and railway routes, 1 with railway and air routes, 29 with only marine route, and 2 with only air route.

As a whole, initial connectivity conditions are ready in countries along the Belt and Road, but still in the starting phase. In summary, the proportion of marine and land connection is the highest, followed by the air and land connection, and the proportion of railway connection is the lowest. At present, the world economic development center of gravity is shifting to emerging economies and developing countries at an accelerated rate, the global value chain and the supply chain are extending to developing countries more quickly, which will put forward new requirements on the international trade shipping system of Belt and Road, the center of gravity of international trade shipping will shift from the traditional pattern dominated by marine shipping to the comprehensive shipping system of marine, railway, air, highway and pipeline and the coordinated interaction depending on multiple modes of transportation.

Since the successful operation of the first regular train from Chongqing to Duisburg on Mar. 19, 2011, container regular trains have been arranged from cities of Chengdu, Zhengzhou, Wuhan and Suzhou to Europe, and the number of regular trains put into operation increased substantially. On June 8, 2016, China Railway Corporation integrated the regular trains to Europe, including Chongqing–Xinjiang–Europe, Chengdu–Europe, Zhengzhou–Europe and Xiamen–Chengdu–Europe regular trains, by using the brand "China–Europe Regular Train" on a unified basis, and now 39 routes have been planned, linking with 12 cities in Europe. By now, a total of over 1,700 regular trains have been operated, with year-on-year increase of 109%, including 572 as two-way regular trains, a year-on-year increase of 116%. The CAAC signed bilateral aviation transport agreements with 61 countries along the Belt and Road, and air flight lines will be extended to 43 countries recently.

In marine shipping, China has planned and started the construction of strategic pivots in Malacca Strait (Malaysia), Indian Ocean land and sea channel (Myanmar), south Indian Ocean channel (Sri Lanka) and in South Pacific, Mediterranean and the Baltic. In recent ten years, the transportation infrastructure connectivity network of China with countries along the Belt and Road has been initially formed, and basically, the Silk Road corridors linking the big transportation channels domestically and extending worldwide overseas have been established.

The reports on the work of the government in all provinces, autonomous regions and municipalities in 2015 mentioned a total of 114 infrastructure investment projects related to the Belt and Road, with total investment amount of RMB 1.04 trillion yuan. The projects were mainly railways, highways and airports, accounting for 68.8% of the total investment, including nearly 500 billion yuan for railway, accounting for almost half, and 123.5 billion yuan for highways and 116.7 billion yuan for airport construction. Connectivity of various transportation means of railway, highway, marine shipping, aviation and pipeline in countries along the lines will get through the blood veins and channels of the Belt and Road, and enable these countries to take off.

3.5.2 "Big Transportation" at the Economic and Industrial Level: Integration of Infrastructure Construction and Industrialization and Urbanization

Behind the massive economic aggregate and dense population of the Belt and Road countries and regions is the seriously lagging industrialization and urbanization. According to the *Blue Paper on Industrialization: Report on Industrialization Progress in Countries along Belt and Road* published by the Industrial Economic Institute of Chinese Academy of Social Sciences, when the 65 countries along the Belt and Road are classified by the stages of pre-industrialization, initial industrialization, medium industrialization, later industrialization and post-industrialization, it is roughly in the inverted trapezoidal form of "1 + 14 + 16 + 32 + 2".

Take Singapore in Southeast Asia and Nepal in South Asia as examples, the per capita GDP is USD 58,523 in Singapore and USD 1,345 in Nepal; the proportion of the primary industry is 0.03% in Singapore and 34.3% in Nepal; the proportion of the added value of manufacturing industry in the added value of all commodity production sectors is 74.6% in Singapore and 12.8% in Nepal; the employment ratio of the primary industry is 1% in Singapore and 66.5% in Nepal; and the population marketization ratio is 100% in Singapore and 18.2% in Nepal. In general, countries along the lines are in different stages of industrialization and urbanization, with quite big differences in geographical location, economic development level, political environment, resource advantage and social customs.

At present, China is actively planning and constructing the six main "economic corridors" and a number of Maritime Silk Road strategic pivots in the Indian Ocean, South Pacific and the Baltic with countries along the Belt and Road. The physical structure of both economic corridors and strategic pivots consists of transportation infrastructures such as railway, highway, flight routes and passenger terminals, cargo stations, airports, harbors, docks, custom house and frontiers.

The completion of transportation infrastructures will greatly speed up the circulation of people, materials, energy and capital, expand the domestic and international material flow and throughput of countries along the lines, promote the import and export trade, enhance market vigor, and increase attraction to domestic and foreign investors, thus promoting the development of local industries. Furthermore, the connectivity of transportation infrastructures can effectively promote the interaction and cooperation between the east, central and west regions along the Belt and Road, inject motive force to the new round of opening up, help more industries in countries along the lines expand in the world, and participate in international competition, to bring greater space of growth and development potential for the economic and social development in these countries.

In the traditional construction of transportation infrastructures in other countries, mainly the simple pattern of infrastructure construction was adopted. Practice has shown that this pattern is not in line with the current phase of economic and social development level in these developing countries. Take the Tanzam Railway as an example. This railway, with a total length of 1860.5 km, was the biggest complete set foreign aid project of China, and also a "golden brand" of aiding Africa. However, the present Tanzam Railway has developed a series of problems such as weak maintenance, chaotic management, frequent accidents and long-term losses, already at the edge of paralysis. The failure in operation of Tanzam Railway has revealed to a certain extent the serious disadvantage of the pattern of pure infrastructure construction.

In the transportation infrastructure construction in countries along the Belt and Road, China should well adopt the TOD (transit-oriented development) pattern in city construction, and the core of this pattern is development guided by transit. On the basis of the TOD concept, the TOD pattern can be further expanded into the "transit+" integrated pattern. With transit taking the lead and putting all into a package, this pattern can well integrate the transportation infrastructure construction with the promotion of industrialization and urbanization construction. To precisely implement the "transit+" integrated pattern, it is necessary to classify in detail the transportation and the transportation based industrial clusters and city clusters.

When classified by transportation, there are forms of railway, highway, aviation, marine shipping and pipeline; when classified by industrial group, there are forms of specialized industrial park, comprehensive industrial park, industrial park close to harbor, industrial park close to airport, industrial park close to railway juncture, and industrial belt along HSR; and when classified by city clusters, there are forms of metropolitans, regional city clusters and national level city clusters. According to the three dimensions of transportation, industry and city, we can obtain the basic operation forms of "transit+" integration pattern. The first is "transit + industry", and its basic operation form combinations are: transit + specialized industrial park, transit + comprehensive industrial park, transit + industrial park close to harbor, transit + industrial park close to airport, transit + industrial part close to railway juncture, transit + industrial zone along the HSR; the second is "transit + city", and its basic operation form combinations are: transit + metropolitan circle, transit + regional city cluster, transit + national level city cluster; and the third is "transit + industry + city", and its basic operation form combinations are: transit + comprehensive industrial park + metropolitan circle, transit + comprehensive industrial park + regional city cluster, transit + industrial park close to airport + industrial zone along the HSR + national level city cluster and so on.

City clusters formed with the HSR as the ribbon and with the stop cities as base points will perhaps become an ideal pattern of new type urbanization. Similar to China, in countries along the Belt and Road, economic development is not balanced between inland and coastal areas, and the poverty-stricken population and backward areas are mainly in inland areas. The inherent characteristics of railways going deep into continent hinterland as a great artery of economic development have determined their absolute position and non-substitutable position in developing the inland economy, it will push the inland area from the edge to the front of opening up, to speed up the urbanization process along the line while quickly integrating the regional industries in surrounding areas and raising the opening-type economic level.

To apply the "transit+" integration pattern along the Belt and Road, the special economic zone construction pattern (Shenzhen and Xiamen), city cluster construction pattern (Yangtze River Delta, Pearl River Delta and Bohai Sea Rim), and construction of national level new districts (Pudong of Shanghai, Binhai of Tianjin and Liangjiang of Chongqing) and other successful industrialization and urbanization construction patterns with typical significance in China can be borrowed.

We should closely combine with the factors of urban population structure, city scale and structure, city cluster development level, industrial structure and industrial space distribution in the countries along the Belt and Road, give full play to the advantages of China in planning, technology, capital, manpower and operation management in the continual upgrading of industrialization and urbanization, help these countries and regions arrange relevant industries and build associated facilities by closely integrating the transportation infrastructure construction with industrialization and urbanization, so as to support the rapid economic development and sustainable development in these countries.

In the current background of low demand, low inflation, low interest rate and low growth of the global economy, we should, based on the transportation infrastructure construction, adopt the "transit+" integration pattern to merge transportation with industrialization and urbanization, build the Chinese overall development version 2.0, erect demonstrative Belt and Road overseas project brands, promote the industrialization and urbanization construction in countries along the Belt and Road, serve the economic development in these countries, and upgrade the regional economic development level, so as to better give play to the advantages of China and demonstrate the characteristics of China.

3.5.3 "Big Transportation" at the Culture and Civilization Level: Marching Towards the "Community with a Shared Future for Mankind"

Since the ancient times, human civilization has spread and developed along with the extension of roads, and all peoples understood and became close to each other as they were linked by roads. The ancient Silk Road two thousand years ago connected many civilization systems on the Asia, Europe and Africa continents and on the sea across the mountains and waters, and promoted the communication, integration and blending of civilizations, writing a brilliant chapter in the history of human civilization.

The Belt and Road Initiative of China is rooted in the traditional "harmony" culture of China. It has activated the Silk Road spirit of thousand years with extensive conclusiveness and general benefiting, carries the common vision of countries and peoples along the lines, and supports and promotes the areas and ethnic groups to consult together outlines, develop together, share prosperity and jointly improve global economic governance, and jointly enhance the welfare of the whole mankind to march towards the "community with a shared future for mankind", with equality as the basis, openness as the orientation, cooperation as the motive force and sharing as the goal.

The concept of "community with a shared future" praises highly the value of "self-discipline and social commitment", "the whole world as one community", "joining in nature" and "man as an integral part of nature"; implements the methodology of "all people enjoying and sharing benefits", "flexible handling according to principles", "combining knowledge with practice" and "being impartial"; advocates the beliefs of "make both yourself and others succeed", "the Benevolent loving the others" and "the Benevolent can serve small nations in the position of a big nation"; pursues the realm of "everything connected with communication between heaven and earth" and "communications achieving harmony"; upholds the principles of "it is precious to achieve harmony", "wide heart embraces all" and "all under heaven are of one family", and emphasizes the co-existence of multi-elements and mutual inclusiveness with "harmony in diversity", with the goal of building a world of universal harmony with "all nations are harmonious in peace" and "warm and cold we share together".

The countries and regions along the Belt and Road have different histories and cultures, different religious beliefs, diversified political systems, complicated geopolitics, and are at different economic development levels, therefore, it is all the more necessary to realize integration and mingling through all-round, multilevel and wide-range communication and exchange. The Chinese government is committed to promoting in-depth people-to-people and cultural exchanges and cooperation with relevant countries and regions through the high-level exchange dialogue mechanism and the high-level consultation mechanism. President Xi Jinping and Premier Li Keqiang have repeatedly emphasized the importance of close people-to-people exchanges and fostering the foundation of friendly public opinion in international exchanges, and vigorously advocated the spirit of silk road and partnership.

At the G20 Hangzhou Summit in September 2016, President Xi Jinping, with the vision and open mind of a big country leader, stressed that "partnership is the most valuable wealth of the G20, and the choice of countries to jointly cope with global challenges". The "Five Consensus" reached at the Hangzhou Summit highly agreed with not only the summit theme, but also President Xi's five proposals put forward in his opening speech, containing the Chinese wisdom of consolidating foundation and win-win cooperation, as the assembly call for the establishment of community with a shared future and shared interests for mankind.

Chapter 4
Strategic Tension for China High-Speed Railway to Go Global

The year of 2016 was the first year of the 13th Five-Year Plan of China, and also the first year of High-Speed railway and mass transit to "go global" in large scale as referred to by the media. The smooth implementation of many projects has not only demonstrated the good situation of China High-Speed Railway (CHSR) going global, but also released powerful kinetic energy to push ahead the Belt and Road Initiative and international production capacity cooperation. However, in this comforting situation, we should think more calmly and remain sober about the challenges and risks for the High-Speed railway (HSR) to go global, so as to assist it to go global more stable and farther.

4.1 Five Tensions for CHSR to Go Global

4.1.1 Geopolitical Tension: Maritime Right and Land Right

CHSR is reputed as the only strategic industry developed by China during the reform and opening up that can change the basic international and domestic political and economic pattern in the whole 21st century. HSR has not only become a diplomatic name card of China, and also become a symbol of new land right daily. Then, will CHSR lead the Eurasian continent into a new land right era? Can it put an end to the maritime era represented by aircraft carrier?

It is known to all that, since the 16th century, maritime power became a focus of attention of world powers. The maritime power is of vital importance to the power of a nation, and the historical humiliation of China in the modern times was also closely related to the fact that the maritime power was passed into the hands of others. Early in 1524, the Spanish conquistador Hernán Cortés concluded that: who that controlled the passage ways between oceans can be the master of the world. The mastery of maritime power not only meant the control of commercial routes and

F. Xu, *The Belt and Road*, https://doi.org/10.1007/978-981-13-1105-5_4

favorable geographic positions, but more decided the ability of a country to export its military forces abroad. Therefore, countries contended intensely with oceans as the center, to get their own maritime right and become a world power.

In fact, the United States is the only big power today able to send its navy forces to all locations of the world and to cut the main "marine traffic routes". However, when the CHSR supports the "land right" with the "right of way" and acquires the symbolic significance of "era of new land right", people have the reason to believe that the era of maritime right supported by ocean and aircraft carriers will surely be changed by the CHSR that has gone global.

The CHSR will go global in three strategic orientations of Eurasia, Central Asia and Transasia. The construction of the Eurasia High-Speed railway can not only change China's long-term dependence on marine shipping in foreign trade, more importantly, it will get through the continental channel to the west, forming a new pattern of opening to the Pacific Ocean.

The "Moscow–Kazan" High-Speed Railway, as an important part of the Eurasian High-Speed transport corridor jointly constructed by China and Russia, will extend further to the east to end in Urumqi of China via Ekaterinburg in Russia and Astana, the capital city of Kazakhstan, and finally connect with the High-Speed railway network in China with "eight longitudinal and eight transversal main lines". This railway is not only designed for a speed of 400 km/h, but also with automatic track switch. It well complies with the ideas of the "Silk Road Economic Belt" initiated by Chinese President Xi Jinping and "Eurasian Economic Alliance" initiated by Russian President Putin, as an important pivot in the "era of new land right".

The Central Asian High-Speed Railway will run through regions with rich resources, just along the "Silk Road Economic Belt", therefore it will play an important role in expanding the economic and trade cooperation of China with Europe and inland African countries, speeding up the formation of the golden trunk line of Silk Road Economic Belt, and pushing the "geopolitics" with the "geoeconomy". The "China-Pakistan Railway" now under construction runs across Pakistan from Kashgar to Gwadar Port, to connect with the west of China.

Kashgar is on the important traffic route from China inland to Central Asia, Middle East, Europe and Africa, and Gwadar Port is close to the throat of Persian Gulf, the important navigation route from Africa and Europe to the East Asia and Pacific Region via the Red Sea, Hormuz Strait and Persian Gulf, therefore the China-Pakistan Railway will not only support the China-Pakistan Economic Corridor, but also greatly change the geopolitical map of the world.

The construction of the Transasian High-Speed Railway will promote the political stability in the Asia-Pacific Region, and help China to get through the marine outfall in the south, forming the new pattern to open to the Indian Ocean, and further enhancing the international geopolitical position of China in Eastern Africa and the whole Eurasia-Africa Continent. The program of Thailand "Rice for HSR" was suspended for a time, but now new change has taken place. The Jakarta—Bandung High-Speed Railway in Indonesia, one of the biggest countries on the Belt and Road, is finally contracted by China after twists and turns.

However, the new land right based on the Eurasian, Central Asian and Transasian railways is not substitution of the maritime right at an appropriate time. First, the practice in the Obama period to treat both China and Russia as strategic opponents will perhaps soon be changed by the new president Trump. People have the reason to speculate that the United States and Russia will possibly cooperate in Europe to combat the "Political Islam", the gradual enhancing of strategic cooperation in combating terrorism will surely produce some negative effect on the connectivity of Belt and Road.

Second, the United States and China will also rival silently on the issue of Taiwan, and the wind will not subside as people wish in the South China Sea, East China Sea, Taiwan Strait and even in the Yellow Sea. During the term of Trump's office, China will not have reduced strategic pressure of marine right on it, instead, such pressure will possibly increase. It is particularly important that, China, as a sea and land compound country, can possibly offset the strategic pressure from the return of the United States to the Asia-Pacific Region as a "strategic offensive" by highlighting the "new land right" significance of CHSR, result in a strategic hedging, however, it can also trigger the strategic doubt in countries along the Belt and Road. Obviously we should be prudent in such land right and maritime right tension resulted from the "offensive realism" based on the Mearsheimer significance.

In the new international political structure, we should establish a new thinking, to change from "either this or that" to "merging each other". We should not only make all efforts to win "land right" on the "belt" of the "Silk Road Economic Belt", but also actively work for "maritime right" on the "road" of the "21st Century Maritime Silk Road", to realize effective connection of "belt" with the "road", and the deep integration of right of way with maritime right.

The China-Pakistan Railway now under construction will not only get through the land passage from China to South Asia, and then to the hinterland of Europe via the Middle East, but also gain access to the Indian Ocean, so that the energy in the Middle East can be directly shipped to Xinjiang of China from Gwadar Port via the China-Pakistan Railway, to fundamentally alleviate the "Malacca Strait Predicament" which is like a fishbone getting stuck in the throat to China. We can say that the China-Pakistan Railway has the typical sample significance of merging the "belt" with the "road" and connecting the "right of way" with the "maritime right".

4.1.2 Economic Tension: Demand and Effective Demand, Small Account and Big Account

According to the prediction by the OECD, the infrastructure construction scale in the world will grow at an annual average rate of 4.9% by the year 2030, and the infrastructure construction investment in the world will reach 12.7 trillion USD in 2020, and 55 trillion USD, accounting for 2.5% of the global GDP every year.

Table 4.1 Planned mileage of railways in regions of the world and investment statistics (excluding China)

Region	Planned mileage (km)		Expected investment (100 m USD)	
	Total	Incl. HSR	Total	Incl. HSR
Southeast Asia	4862	680	689	97
South Asia	23,194	500	129	129
Northeast Asia	3570	1033	103	103
Central Asia	18,761	0	19	0
West Asia	16,183	2000	1174	150
Africa	16,363	3090	615	202
Europe	21,894	6370	1433	860
North America	2863	1232	988	878
Latin America	16,679	0	529	0
Oceania	3448	1748	814	804
Total	127,815	16,653	6492	3222

Source Research report of CRRC Research Institute

According to the global HSR development report issued by the International Union of Railways on November 1, 2016, the mileage of HSR already completed in the whole world was 35,000 km, the that under construction was 15,452 km, that already planned to be constructed was 4246 km, and that covered by long-term plan for construction was 32,065 km, these include 2829 km in Brazil, Mexico, the United States and Canada in America, and 4080 km in Egypt, Morocco and South Africa in Africa. By region, the HSR market in the world is mainly in Asia and Europe, and about 21,760 km will be constructed in the future.

A summary of the railway network plans in all countries shows that the world railway demand in the future 15–20 years (before 2040) is about 120,000–150,000 km, and the HSR construction market developed rapidly, with the demand for HSR at about 15,000–30,000 km (excluding China), and the investment amount expected between 0.6–1 trillion USD (Table 4.1). In terms of market development, HSR will develop rapidly in Asia in short term, and in long-term, the scale in Europe will approach that in Asia. The above statistic data show that the overseas potential HSR market is quite big, and there is still broad development space for CHSR to go global.

However, there are great differences and uncertainties between the huge demand and the effective demand, including many variables, especially, there are many political, social, cultural and even national defense and military variables requiring overall consideration. Take Asia as an example, although 870 km was planned in Iran, 1600 km planned in Vietnam and even 2877 km planned in Thailand, the present development status in these countries shows that even if they have the realistic demand to develop HSR, they are unable to undertake the expensive construction cost.

Furthermore, the complicated relations between the central (federal) government and local governments in some countries and non-government organizations, the ethnic group, religious and cultural factors, and the rivalry between China, the United States, Russia, Japan and EU and regions also influence directly or indirectly the expansion of HSR markets.

Take Europe as another example, although Czech planned 890 km, Poland planned 1127 km, Spain planned 1327 km, France planned 1786 km, and Russia, with its vast land and sparse population, planned 2208 km, the competitors on the HSR market in Europe are already in place, so CHSR can enter the European market only when its standard has been accepted internationally. Besides, China must compete with the powerful opponents in Germany, France and Spain, therefore it will never be easy for China to take a position on the European market.

Therefore between the demand and effective demand, on one hand we must view at the rising of the new round of scientific and technological revolution and industrial revolution, and be fully aware of the boom demand of all countries for HSR on the basis of the strategic considerations of sustainable development, urbanization process, regional common market, construction of comprehensive and green transport system and construction of international passages. On the other hand, we must make an overall view in medium scope the multiple realistic factors of the countries in economy, politics, society, religion, population and security. At the microscopic operation level, we should make rational and prudent evaluation of the actual status of the High-Speed railway target countries on the finance, population density, power supply and economic development level in particular, to acutely identify the demand for High-Speed railway "not really necessary" or "not affordable", and accurately know the effective demand.

In the economic tension, besides the "demand" and "effective demand", the profit and loss of HSR should also be taken into consideration. For the internal affairs of a country, the social benefit of HSR construction is perhaps of the primary importance, because HSR is a public product in itself. However, for the HSR going global, overall, global and long-term taken into consideration should be taken for its strategic significance, profitability as well as social and economic effects.

For the Jakarta—Bandung High-Speed Railway under construction in Indonesia, "the Chinese side will provide the capital and no sovereign credit guarantee is required". What are the risks, profitability and reproducible popularization value of HSR in this "Indonesia pattern"? These questions are worth in-depth analysis and exploration. After all, profiting is the most fundamental and plain appeal. In fact, for Jakarta—Bandung High-Speed Railway, it was originally scheduled to start construction in November 2015, and be completed in 3 years, but by September 2016, land expropriation was only done by 60%. More than that, the agreement also reduced the total project price from the previous 5.5 million USD to 5.135 million USD. The reduction of several hundred million USD means that this HSR line, known as the real "first order of CHSR", is almost profitless in the capital construction phase.

Presently in the whole world, only three HSR lines are profiting: Tokyo—Osaka, Paris—Lyons and Beijing—Shanghai. According to researches by the World Bank, generally speaking, profitability can be realized only when the populations of cities connected by the line reach at least 20 million people. It can be almost concluded with the features of these three lines, especially the economy, distance and populations of the cities they connect, and in conjunction with the initial practice of CHSR going global in recent years that, it can be almost asserted that it will be difficult for overseas HSR projects to be profitable in the predictable future, and they will even bring huge capital pressure and financial risks to the owner entities.

Based on the strategic values of Belt and Road and connectivity, a considerable part of the lines for CHSR going global are planned in regions with sparse population and underdeveloped economy. Of the six main economic corridors of China-Mongolia-Russia, new Asia-Europe Continental Bridge, China-Central Asia-West Asia, China-Indo-China Peninsula, China-Pakistan and Bangladesh-China-India-Myanmar, actually most part is in underdeveloped countries and regions. Therefore, we must face directly the realistic questions of seriously lacking in economic strength, high demand of capital for HSR and poor economic returns in these countries.

The main purpose for CHSR to go global is to shift the trade routes of China from marine shipping to land transport and promote international production capacity cooperation, to change the pattern of world economic trade, and this strategic layout is of far-reaching significance. Therefore, we should not only plan in terms of business, to reckon the "small account" of High-Speed railway going global from the economic point of view and endeavor for the best financial target; more importantly, we should consider the "big account" for HSR on the spillover effect in all aspects, at multiple levels and wide range in politics, diplomacy, economy and society, military and national defense, civilization and culture in the view of the overall concept of Belt and Road, building the "community of shared destiny, interests and value" of the mankind.

4.1.3 Social Tension: Population Circulation and Conflicts Between Civilizations

So far, people have concerned most about the economic benefits and strategic value for the CHSR to go global. But necessary attention is lacking to the social effect of CHSR going global, especially the tension of accelerated population circulation and conflicts between different civilizations.

The modernity connotation of HSR is just like the "time and space detachment" as referred to by the British Sociologist Anthony Giddens. It not only greatly reduces the transport cost of population mobility between regions, and influences the intensity and direction of regional population mobility, but also expands the influence radius of core areas. It can be foreseen that, with the implementation of the strategy of CHSR going global, the population movement along the Belt and Road will surely increase substantially. However, the results of accelerated population mobility are not all benefits, and the "conflicts between civilizations" mentioned by the famous scholar Huntington will possibly become an important strategic issue that the countries along the HSR lines must face squarely with the completion of the railways.

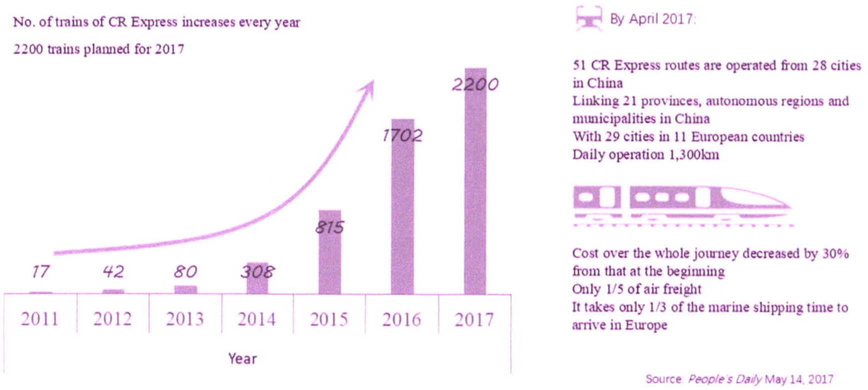

Fig. 4.1 CR express running along belt and road

Presently, in Central Asia, Caucasus, Balkan and the Crescentzone through which the main lines of CHSR going global will run, there are at least two different civilizations, which form the fault zone of civilization where they meet with other civilizations. For example, Central Asia is the fault zone of the five main civilizations of Islam, Slavs, Han and Mongolia, India and Buddhism, in the Balkans, there are eleven main nationalities of Romania, Greece, Turkey, Serbia, Bulgaria, Albania, Croatia, Bosnia, Slovenia, Macedonia and Montenegro, and it is also the intersecting and game playing place of the three major religions of the Orthodox, Catholicism and Islam. Objectively, the existence of these fault zones implies hidden perils and risks of conflicts between civilizations for High-Speed railway to go global.

Today, the issue of "political Islam" as a focus of the world is particularly worth our attention. To China, if this issue is not properly handled in the connectivity of Belt and Road, it will be possible to expose Xinjiang to the front line of conflicts of civilizations. In space, there are large Muslim populations in countries neighboring with and close to Xinjiang, with 633 million Muslims in addition to 14 million in Xinjiang. They include 200 million in India, 180 million in Pakistan, 28 million in Afghanistan, 65 million in the five Central Asian countries, 80 million in Iran and 80 million in Turkey. How will the vast and sparsely populated Xinjiang of China to deal with the possible generalization of "political Islam" in the future? This is a serious issue at hand. If we fail to plan in advance and take positive action to response, we may possibly become passive and even embarrassed, and even the relationship of China with the whole Islamic world could be affected.

China has been quite restrained in "going to the west" in the Belt and Road, especially the "belt", and has not taken action to fill up the power vacuum left after the withdrawal of the United States from the Middle East. This is obviously very wise. To have CHSR go global, our emphasis of construction should be placed on the development of "China Railway Express" (Fig. 4.1), instead of making great efforts to promote the integration of Central Asia and West Asia.

CR Express is the express cargo liner trains running between China and Europe, suitable to freight marshalling trains for containers. In October 2016, the National Development and Reform Commission issued the *CR Express Construction Development Program (2016–2020)*, specifying that CR Express is an important carrier for economic and trade cooperation of China with countries along the Belt and Road and an important grasp to push ahead the Belt and Road Initiative, pointing out the direction for the development of CR Express.

Geographically, three routes of west, central and east were planned for CR Express: the west route runs in the central and western regions of China and exits at Alataw Pass (Khorgos) of Xinjiang, the central route runs across North China and exits at Erenhot of Inner Mongolia, and the east route runs in the southeast coastal regions of China and exits at Manzhouli (or Suifenhe). The domestic parts of all three routes are mostly in the central and western regions. The operation of CR Express has provided rare opportunities for the provinces (autonomous regions) in the central and western regions to change from blocked hinterland to the front for China to open to the west.

In the grand strategic background of CHSR going global, it is the key to success of the Belt and Road to prevent conflicts of civilization between ethnic groups and seek civilized co-existence with countries along the lines for HSR while increasing the intercourse, exchange and merging level of ethnic groups in these countries.

The development tracks of Silk Road in history show that, during the evolution over a thousand years, thanks to the frequent economic and trade relations between the ethnic groups, also accompanied by the mutual understanding gained from cultural exchanges, there was no conflict of considerable scale between ethnic groups, on the contrary, there was a trend of merging and mutual prosperity. The ancient Silk Road played an active role on economic and trade cooperation, cultural exchange and stability of ethnic groups. The Silk Road is not only a road of smooth economic relations and trade, and more a road of cultural goodwill and exchange and road of mutual learning and fusion of civilizations.

In the current situation, innovative thinking is required to create an environment for different cultures to jointly develop, to activate the collective memory of the Silk Road in countries along the lines and inherit and carry on the spirit of ancient Silk Road of "peaceful cooperation, opening and inclusion, mutual learning and win-win mutual benefit". To the developing countries in Central Asia, West Asia and Africa, China must be alert against its own superiority complex or flaunting of strength at a commanding position. We should take initiative to fuse multiple cultures, promote cultural symbiosis, build communities of interests, destiny and responsibilities in political mutual trust, economic fusion, cultural inclusion and civilization mutual learning, to realize joint prosperity of China and countries along the HSR lines.

4.1.4 Psychological Tension: Cognition and Identity

The HSR operation mileage in China has exceeded 22,000 km, however, the cognition of the volume of CHSR has not brought the corresponding international identity. In recent years, the opinion environment about CHSR going abroad is not quite good, and there are prevalent arguments such as "plagiarizing", "vendor's cry at a loss" and "neocolonialism", and a small number of people harboring evil intent even spare no efforts to interpret the Belt and Road policy as continual external expansion of China, to mould the huge image of "Chinese Imperialism". In fact, they intend to create various "theories of China threat", and concocting the "theory of international responsibilities" of China on this basis.

The widely clamoring "theory of China High-Speed Railway plagiarizing" states that CHSR is the "cheap copy" of HSR of other countries. They had the ground that the CHSR development was started with imports from four foreign companies, i.e. CRH1 (Regina C2008 of Bombardier), CRH2 (Sinkansen E2-1000 of Kawasaki Heavy Industries), CRH3 (ICE3 Velaro of Siemens) and CRH5 (Pendolino of Alstom), and all products later were developed on the basis of these types.

In the viewpoint of intellectual property rights, the core of this question is the relation of forward design and reversal design. It is no doubt a wise choice to adopt the imported technical route at the beginning of development, so as to start at a high level like standing on the shoulder of a giant, and this is the so-called reversal design. It is known to all that reversal design is all right as long as one has bought the technology of others and paid the agreed expenses according to contract.

In fact, in the procurement for Beijing-Shanghai High-Speed Railway by the former Ministry of Railway in March 2009, it signed procurement orders of 39.2 billion yuan with the former CSR and CNR, with the subject that the former CSR and CNR will develop by self-reliance CRH380A on the basis of import, digestion and absorption. Although the procurement contract was signed by the Ministry of Railway with the former CSR and CNR, CSR and CNR already paid corresponding expenses to the foreign cooperation partners based on their previous contracts, for example, Siemens got 750 million Euro from this. Why has a matter with so clear fact and reasonable operation been accused, speculated and framed up? Obviously this is related to identity.

On October 5, 2016, at the 3rd anniversary of the Belt and Road Initiative, the railway linking Addis Ababa and Djibouti, the capital cities of Ethiopia and Djibouti, and constructed by China Railway Construction Corporation (CRCC) and China Railway Engineering Corporation (CREC) was officially completed for operation. This is another trans-national railway constructed by China in Africa after the Tanzam Railway, reputed as the "Tanzam Railway of the new period, and it is also the first modernized transnational electrified railway constructed by China in other countries to the complete set of Chinese standards. Not long ago, CRCC signed the design and construction contract for the railway in Zambia from Chipata to Serenjie via Pettauka, China railway is really prosperous in Africa.

For the achievements of China railway on African market, some western media even stated that "the practice of China in Africa is exactly the same as that Britain did more than 140 years ago". What we should pay attention to is that, such groundless accusations of "neocolonialism" have also got support from some non-governmental organizations in Africa. The international opinions, especially the western opinions full of words slandering and demonizing China made us reflect if we are too eager and have overexerted to push the CHSR to go global. It should be known that, there is still a long way from the "knowing to all in our cognition significance to the "hearted identity" of the countries concerned and the general identity by the international society.

The CHSR going global depends on the perfect technical standard, rich operation and management experience and fair market competition, instead of excessively depending on political promotion. Projects excessively depending on political or government promotion may possibly result in strategic suspicion of some countries, and even provide imagination space for conspiracy theorists.

It is natural in practice to endeavor to obtain technical identity in target countries, and more attention should be paid to make concrete efforts to help these countries solve real problems that backward traffic infrastructures seriously restrict the economic and social development, so as to win the value identity by the people in target countries. Value identity is far more important than technical identity. In fact, to the people in target countries, perhaps technology is not a question in itself, and the real question is whether China has been included in the "circle of friends".

There was a very hot post on WeChat about the experience of Fang Xiaoqi and his team touring Pakistan by self-driving in 18 cars. The Pakistan Government arranged military vehicles and rangers to protect the fleet, the people greeted the fleet with flower petals and Cola and Seven-up that they may possibly have not tasted themselves, and babies was sleeping covered by Chinese national flag in the arms of their fathers, waiting for the coming of the Chinese fleet. They got such courteous reception only because "we are Chinese".

In recent years, the construction of China-Pakistan Railway and express highways is undoubtedly first based on the traditional friendship of the peoples of China and Pakistan, instead of the high level of Chinese technology, the good quality of Chinese products and the high performance to price ratio of Chinese projects. For the CHSR to go global, the size of the "circle of friends" will determine how far the HSR lines can go, and only with a big circle of friends can we construct more HSR lines.

4.1.5 Tidal Tension: Globalization and Deglobalization

The development of human civilization in the past 5000 years shows that the future of the mankind will surely be globalization. The "Bretton Woods Agreement" in 1944, the "General Agreement on Tariffs and Trade" in 1947 and the "World Trade Organization" in 1994 broke the trade barriers and economic boundaries between countries. The sciences and technologies in energy, telecom and means of transport, the industrial revolution and electronic information and network revolution have also provided technical support and realization possibility for the globalized development of economy.

In the 21st century, with the rapid growth of international trade, more internationally prevailing standards are formulated, transnational companies rise rapidly, the international financial system get improved daily, countries depend upon each other in economy, politics and culture, and commodities, labor, technologies, capital and information are spreading globally beyond markets and national boundaries. Especially in the recent decade, with the rapid development of modern electronic communication media and the modern means of transportation, and the change in digital technologies such as social media, IOT, big data, financial science and technology and blockchain, virtual reality and artificial intelligence, the time and space distances between people have shortened suddenly, connectivity has become the "meta pattern" in the contemporary times, the whole world has become a "village", and globalization has become an irreversible tide of the times.

However, globalization is not a perfect picture, and the in-depth development of globalization has already resulted in a lot of negative effect. The most outstanding is that the expansion of the gap between the rich and the poor not only exists between poor and rich countries, but even internalized into countries and intensified social contradictions. Therefore in the globalization process, voices of questioning and objection can be always heard. For example, Sammir Amyin, the proposer of dependency theory, believed that "globalization is a reactionary Utopia", and economist Alan Rugma stated that, "globalization has not gone to success, and it has gone to the termination".

According to the global strategist and fashionable lecturer at the TED forum Praag Connor, in the past short period less than 20 years, there were at least four surges questioning the globalization and stating its termination.[1] The first was the September 11 terrorist attack occurring in New York and Washington in 2001, this terrorist attack almost destroyed the trust of the western world on the Arab world, resulting in strengthened boundary control, and giving rise to the surge to question on globalization. The second was the break of WTO "Doha Round Negotiations" in 2006, and some people believed that the global trade will reduce and even wither if globally unified rules are not in place. The third was the international financial crisis in 2008, at that time, exports from all countries dropped, the international credit shrank, and the public policy topics in main countries in the world changed from the neoliberalism form emphasizing the release of market forces to advocating social protection, the "double-edged sword" negative effect of globalization was fermented, and stirred up the deep level internal and external contradictions in economies including western countries. All these were taken by some as the evidences of "deglobalization", the globalization process seemed stopped, and even reversed.

---

[1] [USA] Parag Khanna: Super Map: Rise of Global Supply Chain, Super Cities and New Commercial Civilization, China CITIC Press, 2016.

Starting from 2015, with the FED interest rate increase, the slowdown of economic growth in China, and the reflow of manufacturing industry and automatic production resulted from cheap energy and technological development, the voices of termination of globalization gradually became loud the fourth time. Especially, a series of "black swan" events broke out in the world in 2016, and the globalization process suffered a new serious setback. The first was Brexit by public voting, and resignation of Cameron, the EU, once seen as the model of regional integration, faced unpredictable uncertainty. Then Trump was elected President of the United States, he pursued isolationism and protectionism, and adopted deglobalization policies such as cancelling TPP, levying high tariff on commodities from Mexico and some countries, building the separation barrier along the US-Mexico border, and banning Muslims from entering the United States.

On December 5, 2016, constitution amendment in Italy was rejected by public voting, the prime minister Lenzi resigned, and the biggest opposition party in Italy Five Star Movement was actively seeking a non-binding public voting on exiting the Eurozone. Italy is one of the four main economies of the EU, if Italy exits the Eurozone, it will be basically equal to its exiting the EU, therefore, the EU will face the danger of separation at any time. Furthermore, the EU Parliament declared after voting that it refused to recognize the market economy country position of China, Germany clearly stated opposing TTIP, and the policy balances of more and more developed countries inclined to the trade and investment protectionism side.

This series of many phenomena including "black swan" events showed the rising of worldwide populism and deglobalization movement were rising, the trend of economic nationalism was spreading, and the momentum was strengthening in all countries to care internally. The negotiation processes for high standard regional trade protocols represented by TPP and TTIP are suffering setbacks, and variation may also possibly occur in the mature free trade arrangements represented by NAFTA.

It is certain that serious tension has produced between the irresistible globalization and the populism, nationalism, conservatism, isolationism and deglobalization that are difficult to be contained, reflecting that the world pattern is now at a new cross after the end of the cold war. Conservatism is returning with a strong momentum, people are eager for changes, social splitting and value differentiation are continually deepening, and there are many puzzles and interest games in the world between disorder and governance, separation and combination, and break and establishment. A new international order is in the run-in and collision process, and the world faces a future with more uncertainty. We must draw on advantages and avoid disadvantages for the CHSR to go global in the international situation of intense gaming between globalization and deglobalization.

On one hand, we must have a firm belief in globalization. We should comply with the tide of economic globalization, world polarization, culture diversification and society informatization, and work hard to safeguard the global free trade system and an open world economy. Former UN Secretary General Cofi Annan pointed out profoundly that opposing globalization is like opposing earth gravity. The tide of globalization is irreversible. There have been many problems in the evolution of globalization, and some are even quite serious, but globalization should not be denied because of these problems, and we should not totally deny globalization simply because of problems. Now the question is not to reject globalization, but what kind of globalization we want. We should explore a new road of globalization while "sublating" the existing disadvantages of globalization.

Early at the 29th annual conference of Davos World Economic Forum held in January 1991, the idea of "responsible globalization" was put forth. We should uphold and practice this idea, to turn the undesired globalization of partial economy into "inclusive and responsible globalization", and mould a generally preferential, vigorous and sustainable new type globalization. President Secretary Xi Jinping clearly explained the main idea of "new type globalization" in his keynote speech at 2017 Davos Forum, i.e. realizing rebalance of economic globalization process, correctly selecting the route and rhythm to merge into economic globalization according to the national conditions, paying attention to equity and letting different countries, strata and people share the benefits of economic globalization. In his letter of congratulation to the opening ceremony of 2017 annual conference of Boao Forum for Asia, President Xi Jinping emphasized to jointly promote the more vigorous, more inclusive and more sustainable economic globalization process. The CHSR should adhere to the strategic focus of globalization, and firmly implement the strategy of going global.

On the other hand, in the course of CHSR going global, great efforts should be made to pull the investment and internal demand in countries along the lines by means of infrastructure connectivity, economic and trade cooperation, production capacity cooperation and financial support, to reduce the difference in development level between regions, and realize joint development while upgrading the overall regional development level. We should combine economic globalization with localization, help more countries get rid of poverty and become better off, promote countries along the lines, especially the developing countries and civilized ancient countries to rejuvenate jointly, and build the new type state relationship of "equity and balance, sharing and win-win on regional common market.

In short, China should think and act really on the position of target countries, to closely meet the needs of economic and social development in target countries, instead of merely pushing forward HSR construction. The construction of HSR should be used as a lead, to promote the industrialization and urbanization processes in countries along the lines, improve the people's livelihood and provide more jobs in local areas, and promote the rapid development in economy and society in target countries, so as to bring definite material benefits to the people of target countries, and jointly explore a new and sustainable road for collaborate development of regions.

4.2 The CHSR Going Global: Work Hard to Press Ahead Despite Setbacks

4.2.1 Rare Opportunities

It is in a rare period of important strategic opportunities for the CHSR to go global. In general, the comprehensive strength of China is enhancing constantly, the diplomatic image of the state has substantially raised, the environmental situation of peaceful development is good, and the state leaders of China attached high importance to the internationalization of railway equipment, recommending the CHSR in person at international stage at a number of occasions. This will help forming the situation of government led planning and High-Speed railway enterprises going global in an orderly manner.

The Chinese Government energetically pursued the Belt and Road Initiative and the strategy of CHSR going global. Since the second half of 2013, Premier Li Keqiang has acted as a "salesman" in person to recommend the CHSR in a number of state visits. For foreign delegations visiting China, riding the CHSR has also become a necessary activity basically. In the 1970s, important members of foreign governments visiting Japan would surely ride the "Shinkansan". In 1978 when Deng Xiaoping visited Japan, he marveled when riding the Shinkansan "only feel fast, like pushing us to run. So it is more suitable for us to ride this train. Just suitable for us to ride it".[2] Now more than 30 years have passed. Today, important members of foreign governments visiting China would surely ride the CHSR, what a marvelous change!

As far as foreign environment is concerned, the global economy has further recovered, the construction of HSR in North America has started, the Eurasia Continent economic corridor now plays a prominent role, and there is booming demand for the modernization of railways in Africa. The demand on international railway market (2016–2040) is illustrated in Fig. 4.2.

Correspondingly, the market for mass transit equipment has good prospect. According to relevant report issued by the SCI Consultancy of Germany, during 2016–2020, the whole mass transit market of the world will maintain a steady growth, considering factors such as railway technology upgrading and demand in infrastructure construction in various countries. In the coming five years, a compound annual average growth rate of 2.2% is expected for the mass transit equipment market, including a compound growth rate of 0.8% on the market for new projects and 3.7% for maintenance market; by 2020, the total capacity of mass transit equipment will be about 132.777 billion US dollars (Table 4.2).

As far as the own conditions are concerned, China has accelerated the upgrading of industrial structure, and enhanced the complementarity with the development of relevant industries of the state. Especially, after years of development, the CHSR has obtained achievements that amaze the world, having constructed and operating the

---

[2]The Central Literature Research Center: *Deng Xiaoping*, Central Party Literature Press, 1997, P. 62.

Fig. 4.2 Demand on international railway market (2016–2040)

Table 4.2 Mass transit equipment market capacity in various regions (excluding China) in 2020 (100 m USD)

Market region	For new projects	Maintenance market	Regional total
Europe	161.32	222.89	384.21
Southeast Asia	172.71	145.46	318.17
North America	114.02	129.06	243.08
CIS region	71.52	81.87	153.39
South Asia	57.51	21.78	79.29
West Asia	19.14	13.08	32.22
Latin America	17.24	17.27	34.51
North Africa	6.01	7.91	13.92
Oceania	11.99	12.30	24.29
Southern Africa	11.54	10.12	21.66
Central Asia	11.63	11.40	23.03
Total	654.64	673.14	1327.77

Source CRRC Research Institute, *Research Report on Global Strategic Layout of Mass Transit Equipment Manufacturing Industry*

largest HSR network in the world, having accumulated HSR construction and operation technologies to cope with complicated and diversified geological conditions and climate environment and shared and cross line operation for long distance, high density and different speeds, thus establishing a complete technological system for the CHSR.

The CHSR has not only the comparative advantages of advanced and complete technologies, high safety and reliability, good compatibility, high quality and performance to price ratio, guaranteed product delivery period, good adaptability in construction and mature operation experience, but also the advantage of organic combination of construction and equipment, and the comprehensive advantage of investment and financing support. Therefore, it enjoys good reputation on the international market, with powerful competition advantages.

4.2.2 Variables Should not Be Underestimated

To go global, the CHSR also faces unusual difficulties and serious challenges such as raising huge amount of construction capital, unifying technical standards, coordinating with custom quarantine and safety inspection procedures, unifying construction pace and adapting to transnational operation, and is subjected to the complicated effects of factors such as international political environment, market environment, industrial environment, legal environment, financial environment and cultural environment. The outstanding ones include:

The first is the intense competition with enterprises in developed countries already gained technical advantages in HSR such as Siemens of Germany, Alstom of France, JR East Pass and Kawasaki Heavy Industries of Japan and Bombardier of Canada. As reported by Kyodo News Agency of Japan on March 18, 2014, the four main railway companies of Japan established the "International High Speed Railway Association", to jointly export technologies and equipment of Shinkansen to other countries, promote the High-Speed railway standards of Japan to become internationalized standards, and invited railway companies of the United States and United Kingdom that plan to construct HSR, to help obtaining orders and seize shares on the global railway market; in addition, Japanese Prime Minister Abe promoted "diplomacy with HSR construction", planning to provide the most advanced cryogenic superconductor magnetic levitation technique to HSR program of the United States for free.

The second is that relevant countries have great concerns about their own factors such as geopolitics, national security, resources protection and enterprise interests, and there are many difficulties in the negotiations for cooperation. For example, the website of "The Hindu" published an article, saying "India should say no to CHSR". In fact, to India, all economic and trade cooperation with China has political sensitivity, and the uneasy and negative motion is in place at all times.

Third, technical standards have become technical barrier in international trade. In the background that HSR standards have already been monopolized or used by foreign countries, and European standards have the natural exclusiveness, the core technologies of CHSR can be smoothly applied worldwide only after obtaining international patent and property right protection.

The fourth is that the investment and financing and operation patterns are not quite mature, the BT, BOT, EPC and PPP patterns have their respective advantages and disadvantages, with risks in profiting. The fifth is that in some developing countries, the political situation is not stable, with phenomena of religion conflicts, repelling cultures and corruption, asymmetricity of information and the uncontrollable factors in going global.

4.2.3 Achievements Are Accompanied by Setbacks

In recent years, the going global of China railway, especially High-Speed railway, has been in a good situation. The first High-Speed railway constructed by Chinese enterprise overseas—phase II of Ankara-Istanbul High-Speed Railway in Turkey was successfully opened for operation; and the meter-gauge motor cars made by China for Malaysia with the highest operation speed in the world have been completed and left the production line. The first HSR project of China exported with the whole system and whole industrial chain—Jakarta—Bandung High-Speed Railway in Indonesia has started construction for the pilot section; the first overseas railway project with the whole line based on Chinese standards, using Chinese equipment, to be constructed and operated mainly by China and directly link with China—China—Laos Railway started the implementation phase in December 2015.

The "CR Express" has been put into service officially with a unified brand. China-Thailand Railway and the Serbia Section of Hungary-Serbia Railway have been officially started; China and Russia signed the memorandum of understanding on the development and cooperation and survey and design contract for Moscow-Kasan HSR, China and the United States signed the framework agreement on establishing a joint venture company for "America West Express", and China and Iran signed the commercial contract for the reconstruction for High-Speed for Teheran-Meshed Railway. Positive progress has also been made with the going global urban mass transit projects such as the Orange Line in Lahore of Pakistan, Phase I of Addis Ababa City Light Rail in Ethiopia, and Jiling-Hadong line in Hanoi of Vietnam.

Furthermore, the progress for the HSR project from Malaysia to Singapore is being accelerated, the feasibility study for Delhi—Chennai HSR in India is pushed ahead steadily; and important phase results have been obtained in the basic feasibility study work for the Atlantic-Pacific Railway that will link Brazil and Chile with a length of 3560 km. In October 2016, the railway from Addis Ababa, the capital of Ethiopia to Djibouti was officially open for service. This is the first modern electrified railway constructed by Chinese enterprises overseas by adopting the complete set of Chinese standards and using Chinese equipment the first time, with a full length of 751.7 km and a total investment of about 4 billion US dollars.

The going global CHSR has not only performed well in construction, the achievements in equipment export are more eye-catching. In a short period of a dozen years, while growing into the largest HSR market in the world, China has also become an important supplier of mass transit equipment in the world, having scored extraordinary achievements and development speed beyond expectation. In the meantime, the production capacity cooperation symbolic with the export of HSR equipment has been advanced in full swing, and China has entered institutional production capacity cooperation with nearly 20 countries, and a large number of key projects have been landed.

According to a report on the website of "Frankfurter Allgemeine Zeitung" on May 26, 2015, Haner Gather, a director of German Railway Company, revealed that CHSR locomotives would soon obtain import permit from the railway authority of Germany. In the coming three to five years, China will take an important position in the procurement area for railway locomotives and components in Germany. In fact, CHSR suppliers already have the strength to win contracts in western industrial countries, and their production capacity can meet half of the demand on global locomotive market. Siemens, the traditional supplier of the powerful HSR country Germany has stated that it will actively deal with the new international competitor CRRC.

Despite the good situation for the CHSR to go global, and it has sufficient motive force, strength and potentials, it has also encountered much embarrass in its course to go global. Facing the complicated international environment, intense market competition and unfavorable factors difficult to predict, the CHSR has been advancing really in difficulties in its endeavor to go global. Especially in the recent two years, almost all overseas HSR projects were not smooth, setbacks and even subversive conditions became almost normal.

The China-Myanmar HSR project was stopped; the Mexican government declared canceling the contract only four days after declaring contract award to Chinese enterprise; the proposed "rice for HSR" in Thailand is full of twists and turns; although the HSR project in Indonesia was finally obtained, such a project lacking in government guarantee will test the Chinese project company for its overseas surviving capability in the future; the China-Russia HSR project will face serious challenges in technologies and operation; and CHSR also faces a series of geopolitical and institutional risks when entering the Central and Eastern European markets. As a whole, many negotiations were held for CHSR projects, and cooperation intent was reached with many countries, but only projects in a quite limited number have been landed and started substantial construction, especially, there is no inspiring symbolic and leading ones.

4.3 Ten Major Challenges for CHSR to Go Global

We have come to understand more deeply in the summary and reflection of the CHSR going global that, the CHSR can really go global only when it has finally entered the target countries. Therefore, we must pay high attention to the following ten major challenges.[3]

[3]Fei Xu: *Ten Major Challenges and Strategic Countermeasure of China High-Speed Railway Go Global*, *Frontiers of Academia* No. 14, 2016, pp. 58–78.

4.3.1 *Rationally Cognize the Overseas HSR Market, Objectively Evaluate the Global HSR Capacity and View HSR Beyond It*

HSR is "luxurious" infrastructure in a country, and now it is mainly distributed in East Asia and Western Europe, and China, France, Japan, Germany, Spain and the UK are its core markets. Today, countries with completed or under construction HSR projects include Russia, Poland, Saudi Arabia, the United States, Turkey and Morocco, and India, Thailand and Malaysia are potential markets in the future. There are four high thresholds for HSR construction: one is the investment of huge capital, the second is high population density, the third is sufficient power supply, and the fourth is a certain level of economic development. Countries both meeting these conditions and having the intent to build HSR are quite limited in number.

Furthermore, HSR is also related to many complicated factors of politics, society, culture and even national defense and military aspect. For these reasons, there are not many HSR projects under construction in the world today, and HSR is not necessary or is unaffordable in many countries. Many European and American countries already completed large scale construction of traffic infrastructure before the 1970s, now although some of them have the demand for HSR, they are still weak in economic recovery after the financial crisis, therefore, overseas HSR market is far less broad as imagined. The World Bank thus made the judgment that the biggest HSR market in the world is still China.

How big is the global HSR market? As stated above, the world HSR demand in the coming 20–25 years (before 2040) is about 15,000–30,000 km (excluding China). At present, the UK is implementing the 568 km HS2 project, France plans to build new HSR of 2500 km by 2020, Germany will grant financial support to the railway infrastructure construction already included in plan, and focus on the development of HSR in the modernization construction of railway; Russia has planned 11,000 km of HSR and express railway by 2030, and investment of 52.3 billion US dollars will be first made by 2020, to start the HSR and express railways of Moscow-Kazan, Moscow-Sochi and Moscow-St. Petersburg, with total investment of 4.9 trillion Ruble; India will build a "diamond quadrilateral" HSR network of 10,000 km, and will upgrade and reconstruct existing railways of 65,000 km in the coming 5 years. Besides, Australia, Brazil and South Africa have also put forth HSR construction and development programs in large scale.

Recently, a research prediction published by Boston Consulting Group shows that the total value of global HSR market in 2019 will be about 133 billion US dollars, higher than the 112 billion US dollars in 2014. Furthermore, according to an investigation report of Germany, the global railway market capacity in the future is expected to grow at an annual rate of 3.4%, the predicted annual average growth rate of HSR market capacity is 1.3%, and traditional railway accounts for 3/4 of the total market shares of railway. These figures show that, it is no doubt that the HSR market has a considerable size, but the ordinary railway products in over 200 countries in the world are even more massive, being one or even several dozen times the HSR market.

In the 21st century, with the great development of the new round of scientific and technological revolution and industrial revolution, plus the daily serious problems of global energy crisis, environmental pollution, climate warming and traffic safety, and in the great background of global economy changing to the low-carbon mode, the international railway (mass transit) markets showed an obvious trend of rejuvenation, with passenger transport developing in the direction of High-Speed, urban mass transit and combined transport, and cargo transport in the direction of heavy loading, swift transport and material flow.

Developed countries in Europe and America, Japan and Russia are energetically developing HSR networks and urban mass transit, and more developing countries in Central and Eastern Europe, Asia, Africa and Latin America have also taken priority in developing railways. These countries have adjusted and implemented new strategy for transport development on aspects of sustainable development, progress of urbanization, regional coordinated development, construction of comprehensive transport system and international passages, and increased input to the construction of railway (mass transit).

Therefore, for the CHSR to go global, we must avoid talking about HSR alone, and should view it in the grand background of Chinese railway going global. Furthermore, the "trans-HSR" or express railway between HSR and ordinary railway is multi-unit trains with equipment grade higher than ordinary railway and speed approaching HSR, and it has high performance to price ratio, worth our full attention.

4.3.2 Clearly Know the Diversified Conditions in Target Countries, and Precisely Grasp the Different Demands of Host Countries for HSR

To really bring HSR into target countries, we should clearly know the diversified conditions in target countries, and precisely grasp the different demands of host countries for HSR. Conditions in target countries, especially in countries along the Belt and Road, are diversified and plural. In countries along the Belt and Road, the social systems, economic level, cultural customs, religious briefs, historical traditions, geographic environment and development roads are different, and there are possibly also differences in the track system, construction method, operation pattern and investment financing for railway development.

Take religious belief as an example, in countries along the Belt and Road, several religions coexist, in Central and West Asia and North Africa, Islam is dominating; people in the Central and Eastern Europe mainly believe in Christian and Islam; in Southeast Asia, Buddhism and Islam dominate, with Christian, Confucianism and Taoism coexisting; South Asia is dominated by Hinduism, with Islam, Buddhism and Sikhism coexisting; and the South Pacific region is dominated by Christian.

Another example is the process of industrialization. The countries along the Belt and Road cover all phases of the industrialization, with great gaps in industrialization level between them. The five countries in Central Asia are respectively in the initial period and later period of industrialization; most of the Southeast Asian and South Asian countries are in the early period of industrialization, and Nepal, Bangladesh, Cambodia, Laos and Myanmar are listed by the UN in the least developed countries; most countries in the Central and Eastern Europe, West Asia and Middle East are in the later period of industrialization.

One more example is the track system, which is divided into standard gauge, wide gauge and narrow gauge. China, Iran and Turkey use the standard gauge of 1435 mm, Russia, Kazakhstan and Mongolia use the wide gauge of 1520 mm, India and Pakistan use the wide gauge of 1676 mm, in Africa, Ghana, Congo, Tanzania and Zambia use the narrow gauge of 1067 mm, and Guinea, Ethiopia, Cameroon and most Southeast Asian countries use the narrow gauge of 1000 mm.

The diversified conditions in the countries along the Belt and Road and the differences in their demand for railways have determined that for the CHSR to go global, we must precisely focus on the specific demands of these countries, and make practical, thorough and detailed initial investigation and study. We should fully know the conditions and problems in the host countries, and prepare specific railway plans and integrated technical solutions on survey, construction, equipment, operation, maintenance and investment and financing on the basis of the comprehensive study of the basic conditions of these countries.

We should also provide sufficient detailed data about investment, cost, operation and passenger capacity for decision-making by the host countries. China should make full use of the advantages in the HSR construction data it has and share them with the project countries, so that they can know if it is reasonable to build HSR, how much investment it will require, and what economic benefit it can produce, to help them make better decisions.

4.3.3 Well Combine HSR Construction with Industrialization and Urbanization, to Upgrade the Industrialization and Urbanization with Railways

A HSR is not only a traffic line, but also an economic line, tourism line and a line of people's livelihood, and to specific localities, it is even a life line and a main artery of national economy. Usually an industrial plan is behind a HSR line, and several HSR lines mean the restructuring of the regional economic map and the rising of urbanization. HSR will act as wings for the countries along the Belt and Road to take off in their industrialization and urbanization drive, and bring hopes to the local people for economic revival, especially to countries located in the spine areas of Asia with crisscross roads.

Therefore, we should upgrade the "two tracks" of HSR to industrialization and urbanization, taking foot on HSR and also going beyond it, that is, we should go further from the simple HSR construction to the integrated planning for HSR construction with resources development, industrial development, urban construction and industrial rejuvenation in the areas along the lines, to fully release the "HSR dividend" and give full play to the "driving effect" and "spillover effect" of HSR. This can not only reduce the resistance in building the HSR only and make it more attractive to target countries, but also better improve people's livelihood and create jobs, to promote the development of local society and economy.

The operation of HSR lines will release the cargo transport capacity of existing railways, increase the transport capacity of existing railway network, speed up the movement of people, materials, energy and capital, expand the domestic and international material flow and throughput of host countries, promote the import and export trade, enhance market vigor, and increase attraction to domestic and foreign investors, thus promoting the development of local industries. Furthermore, HSR can effectively promote the interaction and cooperation between the east, central and west regions along the Belt and Road, inject motive force to the new round of opening up, help more industries in host countries to expand in the world, and participate in international competition, to bring greater space of growth and development potential for the economic and social development in the specific countries.

Urbanization is an important embodiment of the "spillover effect" of HSR. At present, the urbanization rate in developed countries has reached 80%, but it is still quite low in countries along the Belt and Road. The HSR is the ribbon for new type urbanization, and will lead the future of urbanization as a trailblazer. Forming city clusters linked by the HSR with the stop cities as the bases will perhaps an ideal pattern of new type urbanization.

Similar to China, in countries along the Belt and Road, economic development is not balanced between inland and coastal areas, and the poverty-stricken population and backward areas are mainly concentrated in inland areas. The inherent characteristics of railways going deep into inland hinterland as a great artery of economic development have determined their absolute position and non-substitutable position in developing the inland economy, it will push the inland area from the edge to the front of opening up, to speed up the urbanization process along the line while quickly integrating the regional industries in surrounding areas and raising the opening-type economic level.

4.3.4 Pay Attention to Promotion of Value Identity, and Build the Community of Shared Interests, Responsibilities and Destiny

During the reform and opening up, Chinese economy maintained High-Speed growth and the overall national strength of China increased continually. China has come to the highlight zone on the world stage as world's second largest economy and the biggest developing country. Every move of China may produce effect on the regional and global economic and political pattern. Other countries have complicated attitudes towards the rapid development of China, they hope to hitchhike with China in economic development, but are also afraid of being controlled by China. In pursuing the Belt and Road Initiative, China will surely touch the traditional pattern of interests, resulting in the discontent and friction from some countries, and even competition and obstruction. In the connectivity of infrastructures, the relevant countries are cooperation partners, and also possibly potential competitors.

For example, Russia worries that the standard gauge tracks used for the "China-Kyrgyzstan-Uzbekistan Railway" cannot be connected with the Russian tracks, and this may further lead to the separation of Central Asian countries from Russia. Especially, in the countries along the Belt and Road, there are great gaps in the economic scale in most of them from that of China, and they have the expectation but also hesitation for connectivity and connection to coastal areas via CHSR. Although China has profound historical origin in the relations with surrounding countries, some countries now see it as potential threat or prevent China as a main competitor, and even try to unite with countries outside the region to impose strategic containment.

Railway is a big artery in a country, and is related to the strategic life line, territory security, social stability and economic development of the country, with major strategic significance. It is also highly sensitive politically and concerned by the public. Normally speaking, neither the government nor ordinary people hope, agree or accept a foreign country to build railway for them, just like in the period of Republic of China, we also boycotted foreigners to build railways in China. This is natural, unless the two countries are really allies. In fact, China has disputes and conflicts of interests with some surrounding countries in territory, island and marine sovereignty, for example, the border problems with India have remained unsettled for decades, the disputes with some Southeast Asian countries for territorial land and sea in South China Sea cannot be dissolved in a short period of time, moreover, as China pursues the Belt and Road Initiative at a high key, some countries along the lines also have misunderstanding and doubt on the original intention of the initiative.

Internationally, for the CHSR going global in recent years, there have been prevalent noises and rampant rumors. Some taunted that it is vendor's cry at a loss as CHSR is not only sold at a low price but also offers financing", some worried "whether China will follow the international standards, take into account the interests of host countries, and implement the social accountability and environmental protection, and whether the HSR going global is a green road or a road of pollution", and some even slandered and vilified China as "seeking hegemony in the world and pursuing for neocolonialism with the Belt and Road Initiative as a tool and the AIIB and Silk Road Foundation as means".

To this, China must clearly state and indicate its position, and accurately convey to the world the Chinese value of "peaceful development, shared development and inclusive development", to enhance the value identify of other countries with China in thinking and concept, so as to fundamentally eliminate the doubts of relevant countries. China should spare no effort to declare to the world that, the Chinese culture advocates harmony, the "harmony" culture of China is long-standing and well-established, entailing the universal view of unity of heaven and man, the international view that all nations live side by side in harmony, the social view that differences are inclusive in harmony and the moral value of good will in people's hearts. China takes into account the interests and concerns of all parties, and seeks interests joining points and the greatest common divisor for cooperation. It hopes "all flowers bloom in the whole garden in spring", instead of "thriving with only one tree". China is ready to share the opportunities in the economic and social development with the ASEAN, Asia and the world, and welcomes all countries to hitchhike the China train of development, either express train or free-riding.

Besides, China should pay more attention to speaking with facts and data. As a big political, economic and responsible country in the world, China is not only playing a constructive role in the international and regional affairs, but also making important contributions to the world economic growth. At present, the world economy is recovering in zigzags. China, as the world's second largest economy, is the bellwether in building an open type world economy, and the development of China will provide more opportunities for the recovery of world economy. The recent research by Prof. Deepak of China and Southeast Asia Research Center of the Jawaharlal Nehru University in India shows that, in 2016, the Chinese economy grew by 6.7%, with contribution to the global economic growth as high as 33.2%. This data is impressive in the background of weak recovery of global economy.

Furthermore, for over three years after China made the Belt and Road Initiative, Chinese enterprises made investment of over 50 billion US dollars in the countries along the lines, a series of major projects have been settled and implemented, driving the economic development in various countries and providing many jobs. China should lose no time to work out the international involving strategy for CHSR, and implement the international railway partnership program, to well combine the national strategy of China with the economic and social development strategies in target countries, and try to quickly build some demon projects in some countries, so that brands can be erected and reputation be raised, and further, we can confront worries with facts, prove our motives with reputation and respond to taunts with multi-win cooperation.

China should further strengthen the good relations with people of all circles in countries along the lines, and mould the positive international image of win-win mutual benefit. "The key to sound relations between states lies in the affinity between their peoples, and this largely stems from mutual understanding", in countries along the lines, China should continually foster large numbers of people who understand, are friendly and close to and ready to help China, to consolidate the social soil for CHSR to go global, build up the foundation of public will, enhance mutual understanding and win value identity.

On this basis, we should let the international society further cognize that the HSR is a tool to make benefits for the ordinary people in all countries, and the envoy to strengthen international exchange and convey civilization. HSR, as a safe, comfortable, convenient, fast, energy saving and environmental-friendly green means of transport, will benefit all countries in the world, promote investment and consumption, and create demand and jobs. China, with its complete HSR industrial chain and advanced technologies, should and can make its contribution in eliminating poverty in the world, enhancing the welfare of the people and realizing the joint development with equity and justice.

In the current external situation and public opinion environment, strategically, the CHSR should be open and inclusive when going global, and we should take initiative to welcome countries in and outside the regions to participate and share the dividend of construction, realize joint development and build a network of common interests. For example, in the cooperation with Central Asian countries, Russia can be invited to join it, and Japan, India and the United States can also be invited in the cooperation with the Southeast Asian countries. Positive interaction with host countries and other countries through competitive cooperation can not only reduce capital pressure, but also enhance mutual political trust and construct new type international relationship.

The 21st century is a century of opening and cooperation, and in the new situation, it is particularly important to strengthen international cooperation on HSR. In design and consultancy, project construction, equipment manufacturing, project supervision and fostering of professionals, the CHSR can obtain competition drive force only through opening, demonstrate its comparative advantage on the international stage only by sharing, realize self-upgrading of weakness only through competitive cooperation, and really find strategic cooperation partners only in a multi-win pattern. Only with this idea can CHSR run about all over the world.

Cooperation in HSR will serve as bridges and ribbons to realize political mutual trust, economic cooperation and cultural exchange between countries in the world. It should be known that the "competitive cooperation", i.e. competition in cooperation and cooperation in competition between countries and enterprises can generate far more benefits than going ahead alone or mere competition, and competition advantages, especially continual competition advantages, are built on the foundation of own advantages combined with advantages in competition with others.

It is the core in promoting value identity to give the local enterprises and ordinary people a real sense of gain. The HSR enterprises of China should really benefit the local people with the project by building high quality projects. They must actively exercise the social accountability of enterprises, and do more good things in the countries where the projects are located. In addition, they must pay attention to operation in localization and operation by local people, local engineering and technical personnel should be recruited as far as possible while ensuring project quality, to grant opportunities for local small and micro enterprises and private-run enterprises to participate and play important roles.

4.3.5 Gaming of Powerful HSR Countries and Breakout of Encirclement

The world powerful HSR countries include Japan, France, Germany, Canada, Spain and Korea, and they are also major competitors for CHSR to go global. Japan is the first country developing High-Speed railway in the world, the Shinkansen High-

Speed rail has the world's most mature commercial operation experience and high safety standards, and no major safety accident has occurred in its operation for 50 years. France is the first country to have High-Speed rail in Europe, and TGV is well known for its High-Speed. Germany has a very high international reputation on the technology of ICE High-Speed rail, especially it can make domestically all the components used for High-Speed rail, being the only country in the world that can do this. The High-Speed railway technical standards of Germany and France have strong influence on the international markets, especially on European markets. Canada's Bombardier is a world class aerospace and train manufacturer, and has long worked in the High-Speed rail industry and is also the architect of the top-rate High-Speed rail.

These powerful High-Speed rail countries all have their own comparative advantages, such as the lightweight and low resistance, disaster prevention and reduction technologies of Kawasaki Heavy Industries of Japan, the active safety, modular car body, quality management system and reliability of Germany Siemens, the ecological design of France Alstom, and the energy—efficiency–economy–ecology (ECO4) of Bombardier of Canada. As developed countries have their respective advantages in HSR technologies, competition and gaming have become daily intense.

On the international HSR market, China has encountered double suppression of "open stratagem" and "conspiracy".

The "open stratagems" include spreading the "China threat theory", open containment and political pressure, and "conspiracies" include sowing dissension and secret denigration. The areas of competition can be Southeast and South Asian countries, as well as Europe, Africa, South America and North America, far from China. The form of competition can be "one-to-one" or "one-to-many". For example, for the Jakarta—Bandung HSR project in Indonesia and Bombay–Ahmedabad HSR project in India, the competition is limited between China and Japan. Examples of multination competition include: on the Myanmar market, China, Japan and Germany are actively promoting their own High-Speed rail technology; China, Japan, France and Spain have all expressed interest in the Singapore-Malaysia High-Speed rail project; the first High-Speed rail project in Mexico attracted 17 companies worldwide, including famous companies such as Japan's Mitsubishi, France's Alstom, Canada's Bombardier and Germany's Siemens.

In the international High-Speed railway competition, some are purely commercial, but more are mixed with complex geopolitical factors. China's competition with Japan on High-Speed rail projects, apart from economic factors, geopolitical factors weigh heavily. Japan's export of the Shinkansen system is one of the strategic cores of the "Abenomics" growth. To some extent, the war of High-Speed railway export has become the "proxy war" between China and Japan for the political influence in Asia and the right to formulate and dominate rules.

In 2015, for the competition in Bombay—Ahmedabad project, the first High-Speed railway in India, the Minister of Land, Infrastructure, Transport and Tourism of Japan made a special trip to India to promote Japan's High-Speed rail "Shinkansen", before that, the JICA experts of Japan even said, "we must get the Southeast Asia HSR even if we win the contract at zero yen". Finally, Japan got the project by offering loans of more than $12 billion at ultra-low interest rate (0.1%) and an extra-long repayment period (50 years). It is clear that this action of Japan was aimed more at politics than economy, trying to "knock it off" with China and prevent the construction of the Silk Road Economic Belt by China.

It also happened that, before Myanmar announced in 2014 the stranding of the China and Myanmar "Kyaukpyu-Kunming" High-Speed rail project, Japan announced at a high profile to offer a free aid of 7.8 billion yen (about RMB 472 million) to Myanmar to help it build railways and other facilities. At that time, Japan Foreign Minister Fumio Kishida who was visiting Myanmar said, "Myanmar is a country at strategic location connecting the ASEAN and South Asian countries, and historically Myanmar has been friendly to Japan, therefore, strengthening the relations between Japan and Myanmar is of great significance". Behind the cancellation of China's plan to build railway in Myanmar, an important cause was the conspiracy of Japan.

One of the main markets for CHSR to going global is Southeast Asia, and Thailand is the first market for the competition between China and Japan on High-Speed rail in Southeast Asia. In December 2014, China and Thailand announced the joint development of High-Speed railway as witnessed by the premiers of the two countries. For this project with construction cost more than $10 billion, loans would be provided by China, later, Thailand stated that the loan interest rate of 2% for capital construction and 4% for operation management offered by China were too high, then Japan put its foot in by the opportunity, and started frequent contacts with Thailand from February 2015. On May 27, 2015, the Japanese government officially signed the memorandum of cooperation with Thailand, for the first High-Speed rail (Chiangmai–Bangkok) in Thailand, the Shinkansen technology of Japan would be imported, and it was a setback of CHSR in the competition for the Southeast Asia market in recent years. The main reason for Thailand's final choice of Japan is that the Japanese government offered a loan at interest rate less than 2% to the High-Speed rail in Thailand by providing the "official development aid" (ODA) loans. Behind the low interest rate is still the in-depth consideration by the Japanese Prime Minister Shinzo Abe to actively push the Shinkansen to "go global" to boost Japan's economy and political power. Japan was eager to get Thailand's High-Speed rail project, because Japan's Shinkansen system, although with 50 years of operating experience at home, had only been exported to one place: Taiwan of China.

The Singapore-Malaysia High-Speed rail is a new battleground between China and Japan's High-Speed rail in Southeast Asia. This High-Speed rail, connecting Kuala Lumpur, the capital of Malaysia and Singapore, is 350 km long and is expected to cost $15 billion. Malaysia announced in April 2016 that Singapore and Malaysia were expected to sign a memorandum on this High-Speed rail project in the middle of the year, and competitive bidding will be arranged in the first quarter of 2017. China, Japan, France and Germany have expressed their intention of competitive bidding or participation. Although a game involving four countries, China and Japan, having fought fiercely on the High-Speed rail projects in Thailand and Indonesia, are regarded as actually the most powerful competitors on this project.

The Japanese bidding consortium was formed by JR East Japan Passenger Railway, Sumitomo, Hitachi and Mitsubishi Heavy Industries; China organized a consortium of six enterprises led by China Railways Corporation, and consisting of CREC and CRRC. Singapore media reported that Singapore was in favor of Japanese companies as they have rich experience in rolling stocks and signaling systems. On the other hand, Malaysia, with most of the route located on its territory and to bear the huge cost, paid more attention to the solution of China with cheap rolling stocks and technologies than Japan and more convenient conditions in terms of financing.

4.3.6 It Is Imperative to Raise the International Recognition of China's HSR Standards

The 21st century is the century of intellectual property, the century of patent technology, and the century of standards and brands. Whoever owns the intellectual property right and patent technology, and standards and brands, has the right of speech, and can dominate the market. China's HSR standards are the most important cornerstone for CHSR to go global. The process to make Chinese standards internationally recognized is the process to export the brand of CHSR, and the process to export the HSR products and technologies of China that has already gained a huge production capacity.

Empirical investigations and researches show that the biggest obstacle for CHSR to go global is that the standards for High-Speed rail are being monopolized by foreign countries; foreign countries mainly adopt European standards, and Chinese standards are not accepted. At present, there is no High-Speed rail constructed completely in conformity with Chinese standards overseas. CHSR must meet European standards to gain access to the European market. Although some standards in Europe have become out of date, to enter the European market, we have to spend lots of time and make great efforts to pass the European certification for all equipment, which would seriously weaken the comparative advantage and competition advantage of CHSR.

It was reported that the products of a Chinese track switch manufacturer have been certified by the quality inspection center of the former Ministry of Railways, which proved that the products are safe and reliable in performance. However, to obtain European certification, it would cost at least 6 million yuan. More capital has to be invested to renovate the die sets, production equipment specification and process flow of the HSR manufacturers of China. The same circumstance occurred in phase II of Ankara–Istanbul railway in Turkey constructed by China Railway Construction Corporation. The project is only 158 km in total length, but it took more than 8 years from winning the contract in 2005 to official opening for operation in July 2014, and this was related to the complicated European standards. European certification is required for all products and equipment of CHSR, including signal, locomotive, rail, cement, rubber gaskets and fasteners, and also including design specification and process flow, even die sets have to be changed to European standards. These will greatly increase the cost of CHSR, completely losing the original advantages in cost.

The international HSR market can accept the European standards, American standards and Japanese standards, why it cannot accept Chinese standards? This actually includes three questions: first, should there be any Chinese standards on High-Speed rail? Second, is there any Chinese standard on High-Speed rail? And third, why are the internal and external causes of the low international acceptance (or refusal) of Chinese standards on High-Speed rail?

The answer to the first question is "certainly yes". Although CHSR was started late, China it a country in which the HSR develops most rapidly, with the longest operation mileage, highest operation speed, largest scale of construction and most complete technologies in the world, able to undertake "package" export from communication signal, engineering works, traction power supply, locomotive and passenger car manufacturing up to operation management.

With the vast territory of China, the HSR construction has also experienced the test of different climate and geological and landform tests, from high and cold areas to subtropical climate, the data and experience obtained from these complex operation environment for HSR have the unique competitiveness on the international market. In particular, China now operates 60% of the world's total HSR lines, really the world's No. 1 HSR power. Therefore, no doubt, China should have its own standards on High-Speed railway.

To the second question "is there any Chinese standard for HSR", the answer is "Yes". A brief review of the history of High-Speed railway in the world reveals that High-Speed rail was originated in Japan, developed in Europe and thrived in China. "The United States has iPhone and China has HSR", and this is a very good evidence of the high degree of acceptance of CHSR by the international society.

In the course from "import" to "taking the lead", China carried out digestion, absorption and re-innovation unceasingly, in the process of domestic High-Speed rail construction, China has formed a complete system of High-Speed rail design, construction, equipment, operation and safety management standards system and railway equipment brand, with independent intellectual property rights on HSR.

There is a lot of experience worth summarizing with such achievements obtained. For example, China adopted an important policy in importing technologies at that time: all parts must be made in China, even for some parts that must be brought abroad as they could not be made in China for a time being, the foreign company can be included in the list of suppliers only when it has formed joint venture with domestic enterprise. It is this important measure that enabled the domestic manufacturers, including the CSR and CNR at that time, to quickly include the relevant technologies into the whole industrial chain of CHSR while quickly upgrading the HSR technologies. Just because of such policies and measures, especially China's strong ability to learn, digest and absorb, China was able to draw lessons and make independent innovation by learning from others, and form a set of HSR technological system and standards with independent intellectual property rights in a relatively short period of time.

There are three main causes of low international recognition of the Chinese standards on High-Speed rail. First, the European and American developed countries have a mature system of High-Speed rail industry, if the High-Speed rail standards and products of China enter the European and American markets, they will surely impact the existing industry, therefore, as self-protection they will not accept and even resist such standards and products with technical barriers.

Second, as China is not an original country of High-Speed railway technology, and has developed a complete set of technologies mainly through import, digestion, absorption, re-innovation and integrated innovation, although we have the full proprietary intellectual property rights for these complete sets of technical standards, after all, they were developed and formed in a short period of time, and it needs time to recognize and accept the Chinese standards.

Furthermore, the High-Speed rail is a part of China's high-end equipment and high technology, as the industrialization in China is still far from maturity, there is still a big gap compared with Europe and America in the overall level of high-end equipment and high technology. It will be difficult to eliminate the impression such as "getting market mainly by low cost advantage, and the Chinese standards are 'modified version' of foreign standards", and this also resulted in low recognition of the Chinese standards on High-Speed rail on the international market.

The main domestic causes of low international recognition for Chinese standards on High-Speed rail are: first, the original innovation capability is not sufficient for the HSR technologies of China, the foundation R&D input is not sufficient, the forward innovation capability is not strong, there are not many core technologies and patents with full proprietary intellectual property rights, and the capability is still to be raised to formulate industrial standard system for the whole industrial chain and whole life cycle.

Second, the HSR industry of China is lacking in a unified and complete standard system on design, manufacturing and certification, the existing Chinese standards on High-Speed rail are not sufficiently normative and in detail, with obvious deficiency when compared with European standards. Although part of the Chinese standards on High-Speed rail is more advanced in technical level than European standards, it still requires improvement and completion in normativity, stringency and accuracy. For example, the European standards for rails specify the physical and chemical compositions, and the proportions of alloys. Furthermore, China does not have strong awareness, broad thinking and powerful action in publicizing its own HSR standards on the international market, and there is even no complete English version of Chinese standards and codes on High-Speed rail today.

Third, standards are closely related to brands. Although the High-Speed railway brand of China has been gradually accepted internationally, its brand characteristics are not distinct compared with those of other countries. The low recognition of brands on market also directly leads to the low international recognition of Chinese standards on High-Speed rail.

Fourth, China still lacks symbolic and demonstrative projects that are built entirely in conformity with Chinese standards on High-Speed rail. Therefore, China needs to face squarely the severe reality of difficult to enter the HSR market in developed countries in Europe and America, find breakthroughs on the international market to accept Chinese standards, directly adopt Chinese standards in countries with ready conditions, and expand the influence with the Chinese standards in fact (de facto standards) starting from specific points.

It will be a wise strategy to start in countries without independent HSR industry of their own, such as developing countries in Central and Eastern Europe, Latin America and Africa. For these countries, the adoption of European standards or Chinese standards has nothing to do with their own interests, and also the decision-making period is short. China can convince them with the advanced HSR already in operation in China that, although the HSR technologies of Europe and America are advanced, the Chinese products have more superiorities, and the performance to price ratio is more competitive, if they are lacking in construction funds, China can also provide loan financing. In this way, it would be easier for them to accept the Chinese standards, technologies and specifications, making Chinese standards the de facto standards.

To our pleasure, in the Jakarta–Bandung High-Speed Railway in Indonesia already under construction, sufficient preparation was done in the initial phase, the project financing, relocation and technical plan have been settled, the host country government offers good cooperation, and it is little disturbed externally, it is the first order for CHSR to go global in all aspects from technical standards, survey and design, project construction, equipment manufacturing, material supply, to the operation management, personnel training, and comprehensive development along the line. It is being constructed as a high quality project on all aspects of project quality, safety assurance, schedule control, investment efficiency, environmental protection, technological innovation, local development, and whole industrial chain development, and will surely become a milestone in the internationalization of Chinese HSR standards.

4.3.7 HSR Demands Huge Amount of Funds, and It Is Necessary to Break Through the Investment and Financing Bottleneck

High-Speed railway has the characteristics of large initial investment, long construction period, low return on investment, slow investment recovery and high risks. From the experience of developed countries in the world, the basic patterns of railway export in Japan, France and Germany are export of technology, contracting projects and export of equipment (MTU), and they rarely participate in operation, to avoid or minimize the business and market risks in railway operation. For example, for the railway between Moscow and Petersburg the HSR MTU's were purchased from Germany; South Korea mainly imported the High-Speed rail technology from France; Taiwan of China imported the High-Speed rail technologies respectively from Japan, Germany and France.

HSR construction requires large amount of investment, and powerful financing is compulsory. However, it is usually difficult for the host country to get all capital needed to build the High-Speed rail, especially the economic development level of countries along the Belt and Road is generally not high, and often they require the contractor to provide project financing. Therefore, it will be quite common to provide capital for the construction of High-Speed railway in foreign countries in a certain period, so financing is also included in the agenda. For example, for Phase II of the Ankara - Istanbul railway in Turkey constructed by China Railway Construction Corporation, the contract amount was $1.27 billion, and the Export-Import Bank of China provided $720 million in loans.

Generally, not only construction entities need bank support, equipment manufacturers also rely on banks. On April 20, 2014, CNR and the Export-Import Bank of China signed a strategic cooperation agreement, for cooperation amount of RMB 30 billion yuan or equivalent US dollar, for a cooperation term of three years, the EIBC lends money to foreign buyers, to support CNR to export HSR products overseas. In fact, the EIBC is a strong supporter to China's construction of High-Speed railways in many overseas projects, such as the projects in Mexico and the Laos.

Similarly, the Bank of China gave energetic support to CSR in the export of technology, equipment and products. For example, the Bank of China Johannesburg Branch issued performance bond and advanced payment bond for Zhuzhou Locomotive Company under the CSR in a total of $410 million for 4 amounts of the contracts of 459 electric locomotives purchased by South Africa, the mass transit vehicle order with the largest amount in a single contract from China ever since.

As a strategy, "providing funds" for HSR construction will be necessary in a certain period of time, even as comparative advantages against the competitors. However, if this method is adopted generally for CHSR to go global, the pressure of capital will be sufficient to overwhelm CHSR. Even if it can withstand it, the risks are extremely high to raise and invest huge amount of capital to construct HSR. Take the China-Russia High-Speed Railway as an example, the total cost for Moscow-Kazan High-Speed Railway is estimated as 1.068 trillion Ruble (over $18 billion), and it requires $100 million for technological justification only. Where can it get this capital? And how can the transport capacity of the railway be met with the population of Russia?

Even so with Russia, the risks will be greater to offer large amount of financing in developing countries with complicated geopolitical situations. Therefore, the "barter" mode has emerged in response to this situation in the HSR export. By "barter", it means exchange for HSR with resources or energy, such as rice, rubber, copper, petroleum and natural gas, this pattern is acceptable by both sides, and the "rice for HSR" of Thailand is a typical example. The governments of China and Thailand reached an agreement after Chinese Premier Li Keqiang's visit to Thailand in October, 2013, China will participate in the HSR project construction in Thailand, and Thailand will pay part of the expenses for the project with agricultural product (rice). Part of the Moscow-Kazan High-Speed Railway project in Russia is also based on "barter", i.e. the Russian side is ready to exchange for the HSR of China with the agreed petroleum and natural gas. Of course, the "barter" pattern still has risks, for example, how to solve it if the exchange cannot be done for any reason? The answer seems that it can only depend on the credit (of the government, bank or commercial credit) of the other party and the mortgage.

The common contracting modes for overseas HSR include BOT (Building-Operation-Transfer) and BOOT (Building-Operation-Ownership-Transfer). For both of them, the income depends on the return after completion of the HSR project, and it's difficult to get profit in a short period of time, the operation and maintenance costs, the traffic flow and competition from other means of transportation can all affect its profiting condition. Usually it takes a fairly long period for HSR operation from loss to breakeven and then to profiting.

Other forms of HSR contracting include BT, EPC, F+EPC and PPP. BT (Building-Transfer) is transfer after completion, without operation. EPC is commonly known as the turn-key overall contracting—"Engineering–Procurement–Construction", the contractor will undertake the engineering, procurement, construction and trial operation service work for the project, and take full responsibilities for the quality, safety, schedule and cost of the contracted project. F+EPC means "Finance +EPC", i.e. the EPC mode plus financing. This mode integrates the project financing and contracting, giving full play to the core role of engineering design on the basis of solving the financing for the project construction. PPP, or the "Public-Private Partnerships" mode, focuses on the operation mode, risk sharing mechanism, allocation of investment and financing functions and project monitoring and evaluation of the cooperative projects in the public and private sectors. The project requires the private sector to be responsible for all the investment for the project, and profit will be realized by charging from users to get back the investment under the supervision by the government. BT, EPC, F+EPC and PPP all involve investment and financing, and it will be difficult to implement any of these modes if investment and financing is not solved.

Among the above modes, BOT is the most commonly used and most mature. However, based on past experience, the economic risks remain huge. For example, the High-Speed railway in Taiwan of China, open for traffic in 2007, is by far the world's largest BOT High-Speed rail, the total length is 345 km, with a total investment of 480 billion NT (about RMB 120 billion yuan), the average cost is equivalent to about RMB 347 million yuan per kilometer. It was constructed and operated by the private Taiwan High-Speed Railway Corporation. Since its operation, Taiwan High-Speed Railway has been losing money due to insufficient transport volume, excessive interest burden and high depreciation amortization, and now it has entered the "bankruptcy countdown" period. Yeh Kuang-shih, the person in charge of the communications department in Taiwan, required either additional allowance of 30 billion NT and extending the concession period of HSR operation from 35 to 70 years, or getting 500 billion NT to accept the bankrupted Taiwan High-Speed Railway.

With BOT, in addition that the return after completion is (obviously) lower than expected, the more serious risk is that no money can be obtained during operation after completion. Although the contractor is allowed to operate it for several decades, the project operation risks in a foreign country are beyond the own control, so the work may be done in vain at last, and it would be better to hand it over directly. Therefore project construction entities prefer the BT mode without operation, money should be paid when the project is handed over, or resources or energy can be used for exchange if capital is not available.

There are several levels for HSR to go global: the first is to export locomotives, the second is to participate in High-Speed railway construction, and the third is to provide financing. These three levels face different risks. The first level generally does not provide financing, and the risk is minimal. The second level has two cases, either providing financing or not providing financing. Financial risk is concentrated at the third level. In actual operation, it should be analyzed according to the actual conditions whether technology export, equipment export, EPC (engineering contracting) or F+EPC (financing + engineering contracting), or BOT, BT, BOOT, PPP mode, or Sino-foreign joint venture should be used. The general guiding ideology is to innovate with the cooperative business mode and investment and financing mode for HSR industry, vitalize the project cash flow, improve enterprise cooperation economy and motive power, strengthen risk prevention and control, and standardize enterprise behavior in an overseas operation.

Generally speaking, for projects with long-term steady operation returns, stable political situation in the host country and with fairly strong economic capacity, the BOT or PPP pattern can be adopted, to share interests and risks; for projects with major strategic value in terms of connectivity, but with poor economic return, high capital demand and serious insufficient economic capacity in the host country, policy support should be offered in the form of Sino-foreign joint venture, to ensure continual operation of the project, of the equipment export, project contracting, financing + EPC method can be adopted, to establish R&D center and perform localized operation by manufacturing enterprise, and at the same time maintenance and project service market should be opened up.

4.3.8 It Is Imperative to Strengthen Top Level Design and Overall Planning

For CHSR to go global, it is necessary to strengthen top level design and overall planning, and it goes without saying in its necessity and importance. However, the real situation is that there is no macroscopic management institution at the national level, no systematic organization for ministries and commissions, domestic financial institutions are lacking in effective integrated management in the overseas financing, and there is no effective coordination between enterprises. It is quite often that enterprises participate in project by their own separately and alone, disorderly competition and vicious competition occurs occasionally among enterprises, often resulting in the awkward situation that conflicts arise between people on their own side.

These not only damage the interests of individual enterprises themselves, but also reduce the opportunities for themselves or joint competition to bid for international major projects. What is more serious is that it will harm the overall interests of the industry and the national strategic interests. The most typical examples are the two bidding projects of CSR and CNR in Turkey and Argentina. Before CSR and CNR were merged into CRRC, out of their respective considerations, these two state-owned rail transit equipment manufacturers did not joint hands to go global, instead, they competed on the overseas market with each other. In the bidding for locomotive project in Turkey in 2011, CSR and CNR forced the prices down each other, CNR bid at a price with almost zero profit, but finally, the order was taken by a Korean company.

In 2012, the Argentina government declared tendering for the urban rail vehicles. In the first round, CNR overwhelmed foreign competitors at the price of $2.39 million/vehicle. When it was about to win the contract, CSR suddenly came into the competition, and offered a price with sharp reduction of nearly 50% from that of CNR. This shocked the Argentina tenderer, which then demanded that the bidding price must not exceed $1.27 million/vehicle. Finally, CSR got the project at the price of $1.21 million/vehicle. It was followed by the incident "CNR suing CSR for competition by price reduction", which triggered heated debates in the public opinions. In fact, the competition between CSR and CNR was reflected not only by horrifying price war, in extreme cases they even denigrate each other in front of international clients. Such vicious competition has not only wasted resources, reduced the profit space, but also greatly damaged the brand image of Chinese enterprises.

In the capital construction field, the two big enterprises of CREC and CRCC and their subordinated companies also fought against each other on overseas market. In the face of the temptations of huge interests on the overseas market, they, previously united as one, clearly began to "mind their own", and sometimes each of them stuck to its own argument and complain against the other. One party would complain that the other tried to seize the order at prices without profit, injuring both others and itself; but the other party would insist that it participated in competition in the market principle, and state that the offered prices were still profitable.

Such cases are typical reflection of lacking in overall planning and uncoordinated pace for the CHSR to go global. In fact, there were very good practices for China to develop international markets in joint efforts, but they were not carried on now. In the past, the Ministry of Railways had project coordination groups overseas, and there was an unwritten rule and tacit understanding that companies of the same category of China should be separated in different project team, for both capital construction and equipment. For example, if CSR was in the China-America project team, CNR must not bid in that team, and must go to China-Pakistan team.

In the world today, there is a clear trend of cooperation in division of work and cooperation with competition between enterprises. Finance, manufacturing and commerce and trade support each other, and collectivization and collective participation in international competition are prevailing. Both theory and practice have proved that enterprises and even industries can complement each other in advantages by means of M&A or strategic alliance, and then they can go to sea by borrowing a boat and climb up by borrowing a ladder. The merger of CSR and CNR is perhaps a good trial. It is hoped that the CRRC formed by merger and other HSR companies and financial institutions of China will speed up integration, strengthen coordination, to form collaboration as an ecological circle, and shoulder the sacred mission that CHSR will, as a vanguard, lead China's high-end equipment manufacturing strategic industries to go to the outside world.

The SASAC of the State Council has always stressed the three essential elements of "good central enterprises": proprietary intellectual property rights, global brands and international competitiveness. CHSR should also aim at the global market, quicken the steps of collectivization to go global, increase the export product pricing, improve the overall profitability on the overseas market, and comprehensively enhance the international competitiveness in the global High-Speed rail market. Also, the guiding and coordination role of the overseas institutions of the Chinese government and industry associations should be brought into play, to raise the bidding success rate for and returns from overseas projects, and avoid disorderly and vicious peer competition overseas.

4.3.9 Internationalized Inter-disciplinary Talents Are Short in China, and Professionals of Railway Technology Management Are Lacking in Countries Along the Route

At present, for CHSR to go global, we are in urgent need for not only engineering, electrical and mechanical professional talents, but also financial, legal and commercial talents; not only survey and design, project management and operation development talents with good mastery of general international contract provisions such as FIDIC and NEC, proficient in professional knowledge and familiar with overseas codes, but also inter-disciplinary talents with international vision, adaptable to overseas working conditions, and ready to work long time overseas, particularly high-end international talents proficient in international trade rules, local religions and culture, international law and regional laws and regulations, and able to conduct effective communication and negotiations with overseas enterprises and governments.

Now, internationalized inter-disciplinary talents are in great shortage, and their training is urgent and must be put on the agenda as soon as possible. Furthermore, in the internationalized development trend of CHSR, there is greater demand for relevant technical personnel and management personnel with higher requirements, so we must lose no time to get ready internationalized inter-disciplinary talents, and recruit personnel with international project management experience, to meet the needs of CHSR going global in large scale and all directions in the future. To meet the urgent need at present, it is imperative to take special actions to enhance the training on project foreign languages, international standards, project management, and FIDIC and NEC certification engineers, to quickly raise the internationalizing quality of personnel to work overseas.

Countries along the Belt and Road mostly have weak foundation in railway construction and management, lacking in local technical management personnel, difficult to support the development of railway undertaking in their own countries. In Tanzam

Railway, the biggest complete set foreign aid project of China, and also a "golden brand" of aiding Africa by China in the 1970s, after 40 years of operation, with the retirement of the older generation of employees having been trained by China, the project has now developed a series of problems such as poor maintenance, chaotic management, frequent accidents and long-term losses, already at the edge of paralysis. This bitter lesson is worth our pondering. Therefore, railway education going global should be pushed concurrently with the CHSR going global, to reserve local railway technical and management professionals for the target countries, and help them to well and properly construct, use and manage railways.

According to incomplete statistics, countries along the Belt and Road are negotiating with China to build railways with mileage totaling more than 16,000 km, according to the railway staffing of 20 people/km, these countries will generate the fostering and training demand for more than 300,000 people. In fact, in Ethiopia alone, more than 2000 local staff members are currently being trained in railway operations, including train attendants, drivers and technical personnel. In a long run, there will be greater demand for fostering and training of rail transit personnel in countries along the Belt and Road.

Also take Ethiopia as an example, it has planned a national railway network consisting of 8 main railway lines. The phase I of railway constructed was the passage way to the sea from the capital Addis Ababa to Djibouti, about 750 km, and it was opened to traffic in October 2016. The phase II construction is scheduled to begin in 2017, with a total of 2000 km, with three new railway lines of Arvash-Voldia, Waldia-Mork and Mojo-Avasa. The far-term plans include the Waldia-Asseta-Tajura, Thurberta-Beidler Railways, and the Addis Ababa light rail phase II, with a total mileage of more than 5000 km. By that time, Ethiopia will have the largest railway network in Africa, with more than 100,000 people as rail transit professionals.

The demand for training rail transit personnel in countries along the Belt and Road is multi-tiered, including the non-academic education demand for railway construction, maintenance and operation management personnel, senior officials of railway system, as well as academic education demand for training high level professionals for railway research institutions or institutions of higher education in target countries.

To go global, CHSR must first have railway education "going global" first. The first is to attract officials and students of target countries to come to China for training; and the second is to help target countries set up railway education and research institutions, and support the localized fostering of railway professional in target countries.

4.3.10 Multiple Realistic and Potential Risks Need to Be Resolved

In recent years, CHSR has accelerated its pace to go global, but at the same time it also faces increasing risks, and the traditional and non-traditional security risks are intertwined, and overlapped with each other, evolving from the past structural risk into a kind of systemic risks. The main risks include geopolitical risk, social risk, security risk, technical risk, and policy, law and operational financial risk.

First, the geopolitical risk. Global political and economic situation is unpredictable, the strategic gaming around the global governance system and global interests pattern is becoming daily more complicated, the overall risk level is on the rise, with more hot, focus and difficult points, and "black swan" events occur frequently. In particular, the countries along the Belt and Road are mostly in a sensitive and complicated social transformation stage, and there are many problems such as internal political disputes, political party struggle, political turmoil and regime changes. Construction of a High-Speed rail projects usually spans several sessions of government, and the variables during this period can greatly increase.

Southeast Asia and Central and Eastern Europe are the key areas for CHSR to go global. The political situations in many countries in these regions are very complicated. Political transitions and civil and local armed forces in Myanmar, and political unrest in Thailand and Bangladesh can all constrain our connectivity with our neighboring countries. If we cannot deeply sense and comprehensively analyze the domestic political environment in host countries, master its movement, it will directly influence the effect and sustainability of "HSR diplomacy" in these regions, and even it is difficult to control the fate of the HSR project itself.

For example, the China-Thailand railway cooperation project has experienced a number of setbacks and changes. In October 2013, China and Thailand signed a memorandum of understanding, and the "rice for HSR" project was launched. After that, the Yingluck Administration stepped down because of political changes in Thailand, and the constitutional court of Thailand ruled that the project was unconstitutional, so the China-Thailand High-Speed Railway project was completely stalled. At the beginning of 2015, Thailand military government began to reconsider the cooperation on China-Thailand Railway, but the cooperation project was changed from High-Speed railway to ordinary railway. In 2015, the two parties started the negotiation process, and nine rounds of talks were held. There were reports "the project will soon start" four times, but actually the start date was postponed time and again. On March 25, 2016, Thailand suddenly announced that it would raise funds to invest in this railway project, the line would be shortened by more than two-thirds, and it would not build route out of the territory.

Another example is the Mexican HSR project "Mexico City—Queretaro" with a total length of 210 km, design speed of 300 km/m and total investment amount of about $4.4 billion, because a Mexican enterprise in the consortium winning the contract was reported as having stakes with the family of the president of Mexico, it triggered the query from the opposition party and many people, then the Mexican government had to declare cancelling the contract awarding.

Second, the social risk. A High-Speed railway project will face a series of social risks in the stages of justification for project approval, before and during its construction and after its completion. In the stage of justification for project approval, the religious beliefs, folk taboo, social conditions and public opinions, opinion orientation and cultural conflict, etc. in the host country can easily trigger negative hypes full of the narrow nationalist color, or even anti-Chinese mood and social unrest. The situation can be particularly acute when ethnic and religious issues are interwoven with political, economic and social issues.

The social and ecological and environmental benefits of High-Speed rail are also important factors to be considered seriously in the stage of justification. The big uncertainty with the schedule and scale of the High-Speed rail in California of the United States is largely due to the thwart by domestic political and environmental activists in the country. The construction of High-Speed railway will involve the immediate interests of the public in land expropriation, relocation and compensation, so it is of vital importance to win the understanding and support of the local public. Because land in foreign countries is mostly private, the cost for land expropriation is high, and land expropriation is a link with many contradictions and conflicts. China and Laos have agreed on a railway construction plan, but it for a time faced the danger that it could not be built as planned because of opposition by local residents on the land for the project.

In the construction of High-Speed railway, due to the strict requirement on the use of local labor by the owner country, it is necessary to pay full attention to the employment of local employees, otherwise it will lead to protests, strikes and other events. When the High-Speed railway project is completed, it will accelerate the migration of people across the regions and promote the interaction between ethnic groups and people of religions, so the high mobility is highly likely to bring about ethnic conflicts and sectarian conflicts.

Third, the security risk. In the context of the world, the areas along the Belt and Road have long been places with frequent terrorist activities and terrorist attacks, in this region, from the Middle East, to Central Asia, South Asia, and then to Southeast Asia, a worldwide notorious "arcuate zone of terrorism" has formed. With the forming of a High-Speed rail network, more convenient traffic could make cross-border problems such as drug trafficking, population smuggling and stealing into other countries more prominent. For economic or political purposes, transnational criminal forces and extremist forces are likely to be saboteurs of channel security. In case of regional situation tension or bad relationship between governments, High-Speed rail is likely to become attack targets by extreme terrorist forces, to greatly harm the surrounding countries, even the personal safety cannot be guaranteed.

The western border areas of China and Central Asia are seriously threatened by international terrorism, ethnic separatism and religious extremism. The risks resulted from the wreak havoc of these three forces not only threaten the life and property security of the people, but also blow the economic structure, social stability, government administration and public confidence in one or a number of countries, and further causing a series of risk such as economic recession, capital flight, running on banks and flee of nationals for sheltering.

Fourth, the technical risk. Countries along the Belt and Road have complicated and varied climate and geological conditions, from the tropics to the high cold areas, from desert and sandstorm to damp and rainy areas, from the coast to the hinterland, from plateau to plain, and from high mountains and deep valleys to rivers and lakes, the national, social and road conditions differ greatly. CHSR going global will surely face a serious of major problems such as project construction, safe operation and disaster prevention and reduction in areas with complicated geological structures and strong seismic zones, the adaptability and reliability of infrastructure and equipment under extreme climate conditions, connectivity of infrastructure and equipment of different gauges and systems, and railway transportation organization under different religious and cultural backgrounds. All these are new challenges to the HSR technologies.

Fifth, policy, law and operation financial risks. CHSR going global will face the severe requirements and access restriction on policy and legal aspects such as target country market access, enterprise qualification, tax policy, sovereign guarantee industrial policy, foreign exchange control, environmental protection, patent, localization, long-term labor contract, and intellectual property rights, as well as risks of serious fluctuation of prices, exchange rate, interest rate and tax rate.

The anti-monopoly investigations in developed markets, transnational mergers and acquisitions, national security examination, foreign investment review and other regulatory requirements and regulations are also barriers that the HSR enterprises of China must get over. Construction and operation in host countries must also face many problems and risks such as severe tendering conditions, pricing out of market conditions, big difference in financing and operation, high construction cost in land expropriation and relocation and labor, the lower economic returns than expected, and difficulty in the overseas replication of the low cost advantage.

Chapter 5
Strategy for CHSR to Go Global

5.1 Strategic Orientations and Lines for CHSR to Go Global

The CHSR going global is an important ribbon to form a new pattern of all-round opening to the outside world, an important backup to fostering and leading the new advantages in international competition, and an important support to the great rejuvenation of the Chinese nation. If China and countries along the Belt and Road can jointly weave a HSR network connecting the Asia, Europe and Africa continents, the past prosperity and glory of the past "Silk Road" can appear again.

5.1.1 Strategic Directions: West, East, North and South

In the past three decades and more and even earlier decades, marine shipping was the most important means of transportation for China to participate in the world economic transactions and trade. However, in the era with the marine right dominated by the United States, the three most important global shipping lines, namely the Malacca Strait, the Suez Canal and the Panama Canal, are firmly controlled by the United States. In recent years, China has come to deeply realize that, if the High-Speed rail can connect the Eurasia Continent, to form a "land right" geopolitical economy, it not only can resolve the huge geopolitical pressure on the "sea right", but also constitute a strong response more directly to the marine supremacy of the United States.

China is planning and constructing the HSR networks running across the continents in the world (Fig. 5.1). We can see on the map of the world that the strategy for CHSR to go global covers all directions: there will be two lines to Europe in the west to reach Paris, a line running along the shore of the Pacific to the east to the United States, a line to the north to link Mosco, Berlin and London, and extending southward via Thailand to Singapore. Especially, the three lines of Eurasian High-Speed

F. Xu, *The Belt and Road*, https://doi.org/10.1007/978-981-13-1105-5_5

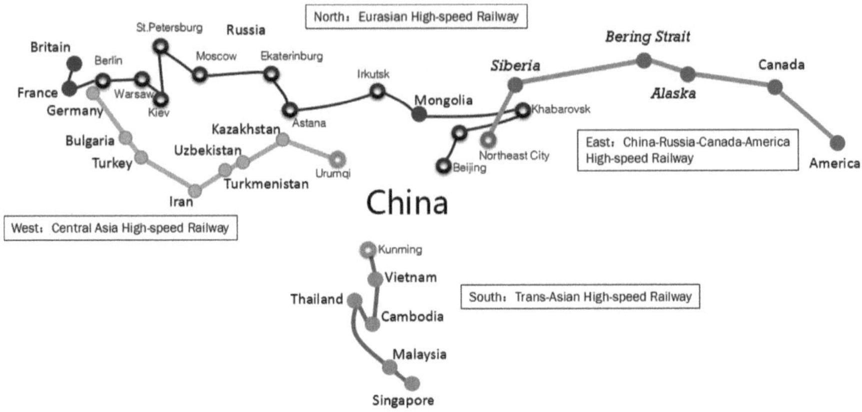

Fig. 5.1 Directions for CHSR to go global: West, East, North and South

Railway, Central Asia High-Speed Railway and Trans-Asian High-Speed Railway have very far-reaching strategic deployment significance. When they are completed, the Eurasia Continent with the super economies of China, EU, Russia and India and the whole African continent will become a huge regional market community linked by High-Speed railways.

In these three strategic directions, the construction of the Eurasian Railway to Europe via Russia will put an end to China's long dependence on marine shipping for foreign trade; the construction of the Central Asia Railway to Germany via Central Asia will expand the economic and trade cooperation with Europe and the inland countries in Africa, to form a gold trunk line for logistics; and the construction of the Trans-Asian High-Speed Railway starting from Kunming, connecting the Southeast Asian countries up to Singapore will get through the port to the sea in the south, to form a new pattern of opening to the Indian Ocean.

5.1.2 Strategic Lines: Six Corridors

According to the above-mentioned strategic directions, China, together with the countries along the Belt and Road, is actively planning and constructing the six main "economic corridors" of new Eurasia Continent Bridge, China–Mongolia–Russia, China–Central Asia–West Asia, China–Indo–China Peninsula, China–Pakistan and Bangladesh–China–India–Myanmar, to build up the land framework for the Silk Road Economic Belt (Fig. 5.2).

New Eurasia Continental Bridge. The new Eurasia Continental Bridge is also called "the second Eurasia Continental Bridge", from Lianyungang of Jiangsu and Rizhao of Shandong in China to Rotterdam Port of Holland, it consists of the railways from Lianyungang to Lanzhou and from Lanzhou to Xinjiang in China, running three

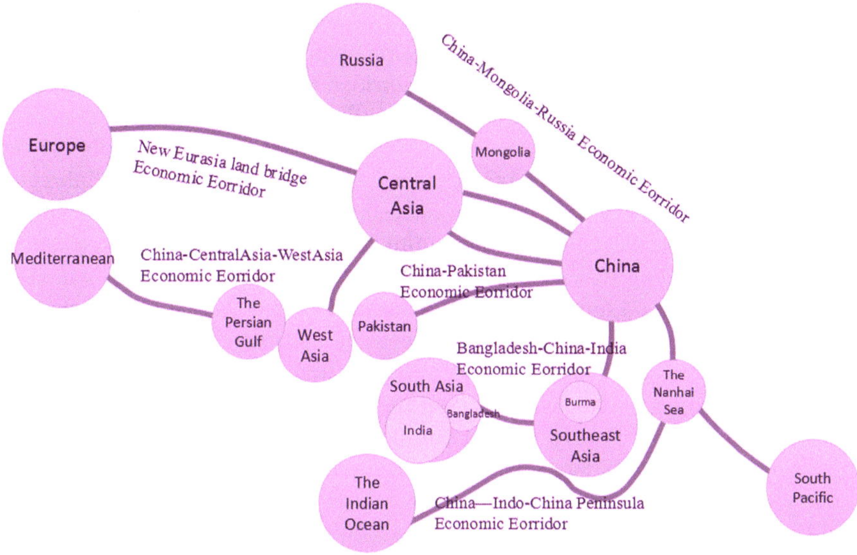

Fig. 5.2 The six economic corridors

seven provinces and regions of Jiangsu, Anhui, Henan, Shaanxi, Ganxu, Qinghai and Xinjiang, then out of China via the Alataw Pass on China–Kazakhstan border, and reaching Rotterdam Port of Holland via three routes. At present, regular trains "Lianyungang–Xinjiang–Asia" and "Lianyungang–Xinjiang–Europe" have been put into operation officially, and the new Eurasia Continental Bridge has become a main logistics channel from China directly to Europe.

China–Mongolia–Russia economic corridor. It consists of two lines: in North China, it starts from Beijing, Tianjin and Hebei to Hohhot, and then to Mongolia and Russia; in Northeast, it runs from Dalian, Shenyang, Changchun, Harbin to Manzhouli and then to Chita of Russia. The construction of "Silk Road Economic Belt" in China is closely connected with the renovation of trans-Eurasia railway in Russia and the "Road of Grassland" initiative of Mongolia, and they will strengthen the connectivity of railways and highways, facilitate custom clearance and transportation, and promote cooperation in cross-border transportation. At present, on this economic corridor, "China–Russia–Europe" railway international freight regular trains including "Tianjin–Manzhouli–Europe", "Jiangsu–Manzhouli–Europe", "Guangdong–Manzhouli–Europe" and "Shenyang–Manzhouli–Europe" are in operation, and basically normal operation has been realized.

China–Central Asia–West Asia economic corridor. Starting from Xinjiang of China, it reaches the Persian Gulf, the Mediterranean and the Arabian Peninsula, mainly covering five Central Asian countries (Kazakhstan, Kyrgyzstan, Tajikistan, Uzbekistan and Turkmenistan), Iran and Turkey. The construction of "Silk Road Economic Belt" in China is closely linked with the national development strategies including the "Bright Road" of Kazakhstan, the "energy, transport and grain" rejuvenating strategies of Tajikistan and the "Prosperous and Happy Era" of Turkmenistan, to deepen the regional international cooperation on production capacity.

China–Indo–China Peninsula economic corridor. China has planned to construct an all-directional traffic and transport network of the east, central and west lines of Trans-Asian Railway with the five countries on the Indo–China Peninsula (Myanmar, Laos, Vietnam, Thailand and Cambodia) with the Greater Mekong Subregion economic cooperation mechanism, to eliminate the bottleneck of lagged traffic infrastructure, and promote the regional economic integration process. At present, countries in the Greater Mekong River Basin are building 9 cross-border highways in E–W and S–N directions, and some of them have been completed. The S–N highway from Kunming of China to Singapore has got through, and the E–W highway linking Myanmar, Thailand and Vietnam is under construction.

China–Pakistan economic corridor. Starting from Kashgar in Xinjiang of China, and ending at Gwadar Port in Pakistan, it has a total length of 3000 km, running through the key hubs of the S–N silk road, linking the silk road economic belt in the north and the 21st-Century Maritime Silk Road in the south, and is a trade corridor including highway, railway, oil and gas conveying pipelines and optical cable channels. This economic corridor is an important embodiment of linking China's the Belt and Road Initiative with the development strategy of Pakistan to boost domestic economy, a flagship and sample project of the Belt and Road Initiative, and a model of China building a community with shared future with its surrounding countries. Both countries have agreed to build up the "1+4" cooperation pattern with the China–Pakistan economic corridor as the lead, the Gwadar Port, energy, traffic infrastructure, industrial park and production capacity cooperation as the main contents, and the AIIB and Silk Road Fund as the support.

Bangladesh–China–India–Myanmar economic corridor. It covers the areas of the southwest of China, east India, Myanmar and Bangladesh, and is a land channel linking the Pacific with the Indian Ocean. The connectivity of infrastructures of railways, highways and communication in this area will speed up the industrial cooperation and trade increase in the region, and promote the linked development of the three main economic plates in the East Asia, Southeast Asia and South Asia.

On the basis of planning the above six main economic corridors, the Ministry of Transport of China has also planned 9 Belt and Road key transportation projects (channels) including China–Laos–Thailand, China–Mongolia, China–Russia, China–Pakistan, China–Kyrgyzstan–Uzbekistan, China–Kazakhstan, China–Tajikistan–Afghanistan–Iran, China–India and China–Vietnam. These nine channels and six corridors have basically formed the Silk Road corridors linking the main transportation channels internally and ramifying globally externally (Fig. 5.3).

Today, the going global of CHSR is following closely the construction pace of Belt and Road, in conjunction with the construction of the six main economic corridors and the nine channels, it will push forward the construction of railway networks to get through the subregions in Asia and between Asia, Europe and Africa, and bring into play the foundation role of railway in promoting the regional resources movement, trading, humanistic exchange and integration of regional economy.

Fig. 5.3 Panoramic schematic diagram of belt and road

5.2 Strategic Principles, Arrangement and Routes for CHSR to Go Global

5.2.1 Strategic Principles

The going global of CHSR requires us to consider the situation and plan strategy, and adhere to the following four basic principles.

First, leading by the government and collaboration with general planning.

HSR is a strategic product with public, social, basic and supportive nature, instead of an ordinary commodity (product). HSR is important facility of a country, and is related to the national strategic lifeblood, territory security, economic development and social stability, especially it has a strong geopolitical color, therefore the decision-making on HSR is surely a state behavior and government will. Also, HSR is a super project rather than ordinary one, with the features of large investment, long construction period and low returns, involving wide aspects and it is difficult to rely solely on market, and it needs the government leading and promotion, coordinating multiple factor variables, making proper top-level design, and strengthening multi-dimensional cooperation and collaboration at the international level, national level, social level and market level.

To enable the CHSR to go global, the government needs to play a dominating role internationally and a leading role domestically. In the overseas consultation and communication with target countries, the government should closely link the HSR going global with the implementation of Belt and Road Initiative and the diplomatic strategy of the state, take into comprehensive consideration multiple factors in politics, diplomacy, national defense and regional economy, adhere to mutual benefit and seeking common points while reserving difference, pay attention to the interests and concerns of all parties in the target (host) countries, seek for the common points of interests and the greatest common divisor for cooperation, and decide at a strategic level the major issues on the target, time and method to go to specific target countries.

Domestically, the government needs to play the guiding, leading and promoting role. The government should steer the direction, erect platforms and make overall planning and coordination, and establish complete mechanism for division of work and cooperation, and clearly divide the responsibilities and functions between the state and society and between the government and enterprises. Of course, the leading role of the state and government should not evolve into administrative intervention.

Second, market operation with enterprises as the main players.

We should give full play to the decisive role of market in resource allocation, and fully mobilize the subjective initiative and operation autonomy of all social sectors and railway enterprises, to fully demonstrate the positions of various enterprises as main players. The society and enterprises should properly follow the market mechanism in their communication, investment and business activities. In the initial period of a HSR project overseas, mainly the government should play its dominating role, however, once the project has been approved and determined, enterprises should play the leading role, weaken the political and governmental factors, and operate in strict accordance with market rules, business principles and prevailing international rules.

Also, they should open up new ways with innovation in the engineering design, construction, equipment manufacturing, operation and maintenance, cooperation form, risk prevention and control, and investment and financing mode, and take initiative actions in aspects of project quality, safety assurance, schedule control, investment efficiency, environmental protection, technological innovation, local development, whole industry chain development, to exercise their rights, discharge their responsibilities, give play to their strong points and share risks and benefits.

Third, consultancy going ahead for precise decision-making.

Consultancy going ahead means that we should acquire an in-depth understanding of the host country demand, link with the host country strategy and planning, justify on the project feasibility, deepen the technology scheme comparison, and settle the financing for the project on the basis of in-depth research of the host country resources, geography, industry, policy, law, etc. to tailor-make the project scheme for the host country and provide comprehensive support from all aspects of technology, equipment, design, manufacture, operation and financing, and make every effort to realize accurate implementation with a specific plan and policy for each country (or even each project).

With the piloting and leading by the high-end design and consultancy, we can better access into the market, find and reserve target projects, and lay a solid foundation for the EPC or the right to say in the project, so that the design and consultancy service going ahead can drive the systematic development of survey, supervision, project management, construction, equipment manufacturing, operation and maintenance as the whole industrial chain, whole supply chain and whole value chain, to further realize the going global of CHSR as a whole in design, construction, equipment, operation and maintenance, technologies, standards and brands.

Fourth, joint construction by multiple parties and industry union.

HSR is a super project with long industrial chain, and requires industry union for joint construction by multiple parties. There are various ways for joint construction, enterprises of railway industry at different value chain positions, and enterprises of different sectors such as railway, finance and insurance can, by joining stocks or other agreements, form alliances as vertical alliance and horizontal alliance, cross-sector joint venture, mutual stock holding, mergers and acquisitions, technology development alliance, the entrusting party and contractor (OEM) agreement, cooperative production alliance, service alliance, franchise, outsourcing or multi-level cooperation, to realize sharing of resources and risks, cost apportionment, mutual supplement of advantages, market expansion, industrial collaboration and integrated construction and operation.

"Multi-parties" include both Chinese and foreign parties. By forming alliances, Chinese enterprises and institutions can go into deep strategic cooperation, not only forming a powerful resultant force, but also avoiding vicious peer competition. The Chinese and high-quality foreign enterprises can join hands at appropriate time to develop local and third party markets, this can make full use of various resources, conduct strategic cooperation with relevant enterprises, play a positive role in such aspects as getting through technical barriers, promoting government negotiations, solving market disputes, and shortening the process to exploit overseas market, shape a good situation of risk-sharing and win-win mutual benefiting, so as to form industry association on the international market, and achieve global allocation of resources.

5.2.2 Strategic Arrangement: Space Arrangement

First, Southeast Asia, South Asia and North Asia.

China is located at the natural center of the Asia–Pacific Economic Zone, by the world's largest continent and ocean, on the west coast of the Pacific Ocean and East Asia, with a long mid-latitude coastline, and borders with 14 countries on land (Fig. 5.4). At present, the geopolitical situation faced by China is becoming daily stabilized, with growing trade relations and closer political and economic cooperation with the surrounding areas.

Southeast Asia, South Asia and North Asia are the main areas for the CHSR to go global for connectivity. In November 2014, at the dialogue on strengthening connectivity partnership, President Xi Jinping advocated again to jointly develop connectivity with surrounding countries, and arrange traffic infrastructure projects in priority in Belt and Road with traffic infrastructure as the breakthrough, to achieve results early in the connectivity in Asia. At present, the China–Pakistan Railway from Kashgar of Xinjiang via Islamabad to Karachi and Gwadar Port is under construction.

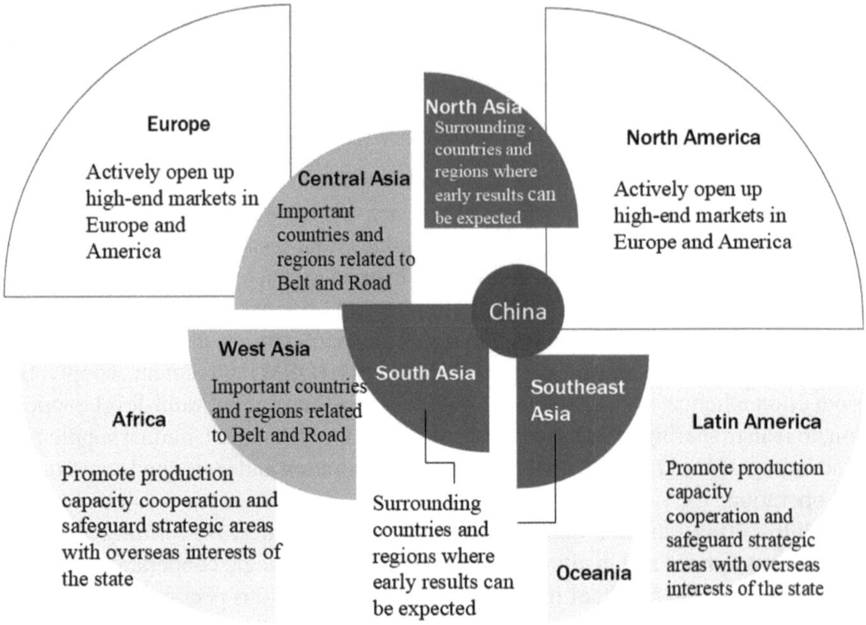

Fig. 5.4 Strategic arrangement for CHSR to go global: space arrangement

We should lose no time in planning the "Trans-Asian High-Speed Railway" starting from Kunming to Singapore via Cambodia, Thailand and Malaysia. Planning should be started for the line which links Pakistan, India, Bangladesh, Myanmar and Thailand, and enters Yunnan of China, and the "south route" which connects Vietnam, Thailand and Myanmar and extends to Singapore via Malaysia. Sections not yet constructed by now include that between India and Myanmar, that between Myanmar and Thailand and that from Thailand to Yunnan of China.

It is expected that by 2020, there will be 11 key projects for China railway to go global in Southeast Asia, South Asia and North Asia, including HSR, with a total length of 7377 km, for total investment of $98.9 billion. In the recent period, we should speed up the China–Laos railway and Jakarta–Bandung High-Speed Railway in Indonesia now in construction, and push forward the projects of China–Thailand railway, Malaysia–Singapore railway, Moscow–Kazan HSR in Russia and east railway in Malaysia, to form the Trans-Asian Railway network as early as possible.

Second, Central Asia and West Asia.

Central Asia and West Asia are main areas for implementing the Belt and Road Initiative by the going global of Chinese railway. The Chinese Government long ago realized the importance of developing the resources in Central Asia and promoting the Central Asian area to develop their economy, and started many years ago the plan to build a HSR crossing the Eurasia Continent via Central Asia and West Asia. Such a High-Speed rail is an irresistible attraction to China, both commercially and strategically. Upon its completion, the crude oil from the Middle East to China and the products made in China for export to Europe can reach the destinations via land without marine shipping. It will provide great convenience for China to import crude oil and minerals from the Central Asia and Middle East, and enable export commodities from China to reach the Middle East and European markets more quickly and conveniently.

Therefore, great efforts should be made to link up the whole railway line and renovate the existing sections, to lay a good foundation for the construction of the "Silk Road Economic Belt" and the Eurasia Land Bridge. In the Belt and Road Initiative, it is planned to build a major channel exiting from Alataw Pass-Khorgos, running through the Central Asia, West Asia and Russia, and finally reaching Europe to join with the railway system in Europe. The planned "Central Asia High-Speed Railway" will start from Kashgar of Xinjiang, cross Kazakhstan, Uzbekistan, Turkmenistan, Iran, Turkey and other countries, enter Europe via Bulgaria and end in Germany. This HSR network is expected to cover 17 countries.

We planned the "S–N corridor" that links the northern Europe and Persian Gulf. The trunk line of this corridor begins in Helsinki, Finland, crosses Russia to reach the Caspian Sea, and then it is divided into three branches: the west route runs via Azerbaijan and Armenia into western Iran; the middle route starts with train ferry across the Caspian Sea to Iran; and the east route runs through Kazakhstan, Uzbekistan and Turkmenistan to eastern Iran. The three lines will meet in Tehran, the capital of Iran, and finally arrive at Bandar Abbas.

It is expected that by 2020, there will be 3 key projects for China railway to go global in Central Asia and West Asia, with a total length of 1840 km, for total investment of $13.9 billion. By 2020, we will complete the electrification transformation of Tehran–Mashhade Railway in Iran, and actively promote the construction of the Tehran–Isfahan High–Speed Railway and the implementation of the initial period work for the China-Kyrgyzstan–Uzbekistan Railway.

Third, Africa and Latin America.

Africa and Latin America are strategic regions for China to promote international production capacity cooperation in railway, expand international space and safeguard its overseas interests. This judgment is based on three reasons: first, they are highly associated with China in foreign trade; second, there are strategic resources lacking in China; and third, the political and diplomatic relations are close, with traditional friendship, conducive to China's expansion of international economic and trade space.

Two routes can be planned for railway construction from the Middle East to Africa: from the United Arab Emirates (Dubai), Saudi Arabia, Yemen to Ethiopia in Africa; or from Kuwait, Iraq, Syria (Damascus), Jordan via Suez Canal of Egypt to Africa. It is expected that by 2020, there will be 13 key projects for Chinese railway in Africa and Latin America. They will include 10 in Africa and 3 in Latin America, with a total length of about 14,421 km, and investment of about $93.55 billion.

In recent period, we should participate in the construction and operation of the "four longitudinal and six transversal" railway lines in Africa, work out and implement the *China–Africa Cooperative Action Plan on Railway (2016–2020)*. We should speed up the construction of Mombasa–Nairobi Railway in Kenya and the renovation of the Belgrano freight railway in Argentina, complete the Addis Ababa–Djibouti Railway, promote the construction of Kenya–Naivasha–Malabar Railway and the High-Speed rail project in Egypt, and actively promote the reconstruction project of Tanzania–Zambia Railway, the Morocco High–Speed Railway and coastal railway project in Nigeria. Complete the basic feasibility study for the Brazil–Peru Railway and Argentina–Chile Tunnel projects that link the two oceans.

The Brazil–Peru Railway will be the transnational construction project most expected by Latin American countries, it will run across the South American continent from Brazil to Peru to link the Pacific Ocean and Atlantic Ocean. In Brazil, it will start from the Atlantic coast on the east at the important town in the southeast of Brazil–Rio De Janeiro, extend to the north–west to cross the Minas Gerais State, Goias State that are rich in iron ore, and then through one of the core agricultural areas with abundant produces of beef and soybean–Mato Grosso State, Porto Velho in Rondonia State and Amazon State, finally enter Peru from the Acre State at Brazilian border, and reaches the sea at the Callao Port, Mollendo Port or Arica Port of Peru.

This planned railway from Pacific Ocean to Atlantic Ocean has a total length of about 5000 km, with about 2000 km on the existing railway system, and new railway will be built for 3000 km. The budget for this project will probably be as high as $60 billion. China, Brazil and Peru will invite bids internationally, and endeavor to cooperate with World Bank and Brics Development Bank to raise funds for the construction of this railway. This railway can break the present monopoly of the international logistics by the Panama Canal under the control by the United States. Moreover, the construction of the railway network will be of important significance to the economic development in the Latin American countries, and China can also export railway equipment and technologies, to shore up its export trade. The joint construction of the railway linking the Pacific Ocean and Atlantic Ocean by China, Brazil and Peru will be an action of win-win mutual benefit.

Fourth, Europe, North America and Oceania.

Europe is one of the cradles of High-Speed rail, and Germany, France, Italy, Spain, Britain, Belgium and the Netherlands already have High-Speed rail lines in operation. By the end of 2011, dedicated High-Speed rail lines with a total of 6830 km (at speed of over 250 km/h) were well developed in the EU, and this High-Speed rail network connects Paris, Brussels, Cologne, Amsterdam and London, the main cities of the five countries in northwestern Europe, forming a complete international network.

At the international level in Europe, High-Speed rail has been deeply rooted in the idea of people, High-Speed railway is regarded as the core content to realize a common transport policy in Europe,[1] and has become the motive force of development in some countries, where High-Speed rail network skeleton has already been

[1]European Commission (Directorate General for Mobility and Transport), "High-Speed Europe: A Sustainable Link Between Citizens", Luxembourg: Publications Office of the European Union, 2010.

completed. The economic crisis has led to tighter scrutiny of some big projects, to confirm the expected benefits from these projects. The main obstacle to High-Speed rail construction in Europe is capital, as the return on its investment is not attractive even when the broad economic and social impact of High-Speed rail construction is taken into account.[2] The EU's funding for European transport network construction is largely confined to research rather than construction.

There are differences between European and Chinese High-Speed rail standards, making the market access difficult. The United States has the largest railway network in the world, but its development of High-Speed railway has lagged behind in recent years. Now the United States, Britain and other countries are actively promoting High-Speed rail layout, to go into Europe and the United States, CHSR should actively carry out high-end product research and development, strengthening product customization, personalization and adaptability, pay attention to energy conservation and environmental protection, to meet the demands of European and American high-end market. At the same time, the planning of "Eurasian High-Speed Railway" should be accelerated, this line will start from London, run through Paris, Berlin, Warsaw, Kiev and Moscow to Urumqi in China, and finally end at Beijing. Preceding research will be made for the China–Russia–Canada–United States High-Speed Railway, which will start from the northeast of China, run through the Siberia to the Bering Strait, cross the Pacific Ocean with a tunnel across the ocean to Alaska, then from Alaska to Canada, and finally to the United States (Fig. 5.1).

It is expected that by 2020, there will be 6 key projects for China railway to go global in Europe and the United States, with a total length of 1840 km, for total investment of $97.34 billion. In the near future, we should accelerate the construction of Hungary–Serbia Railway, which is the flagship project of the China–Europe Railway. We will promote the cooperation on the Baltic Joint Railway, the British High–Speed Railway and the Swedish High-Speed Rail Projects, and also study the construction of High-Speed rail in California of the United States and in Ontario of Canada.

In recent years, the Australian government has been fully aware of the feasibility and necessity of building High-Speed rail and plans to build a High-Speed rail on the east coast and a rail network linking Brisbane, Sydney and Melbourne. The speed of proposed High-Speed train will be over 250 km per hour, so it takes only 2 h and 40 min to travel from Sydney to Melbourne, same as and even quicker than airplane. Research shows that for the first section of the High-Speed rail from Sydney to Melbourne, return of $2.15 will be generated for Oceania from every US dollar invested. If we can act actively to seize the opportunity to build High-Speed rail in Australia and win the right to build it, it will be a great example of CHSR going global.

---

[2]Roger Vickerman, Wang Jiao'e, Jiao Jingjuan and Jin Fengjun: *The Development History and Economic Effects of High-Speed Railway in Europe, World Regional Studies* No. 3 of Vol. 22, September 2013, p. 41–45.

5.2.3 *Strategic Arrangement: Time Echelon*

Planning can be made at three time echelons on the basis of the above-mentioned space arrangement (Fig. 5.5).

The first echelon: quickly get through the route to the south and connect the African continent. Priority should be given to the Trans-Asian Railway, China–Pakistan Railway and Africa Railway. For Trans-Asian Railway, China is neighboring with the Southeast Asian countries, the China–ASEAN economic cooperation, the construction of Maritime Silk Road and the security of South China Sea have fully reflected the importance and urgency of this cross-border railway; in the South Asian countries bordering with China, the industries are highly complementary, there is no barrier of technical standards, and understanding of cooperation has been reached with relevant governments. The construction of the domestic section of the Trans-Asian Railway has been started, and the route to the south will soon be established.

For the China–Pakistan Railway, the situation is pressing as 80% of the petroleum imported to China must go through the Malacca Straight over a long roundabout route, and under the control by others. China–Pakistan Railway is over 2000 km in total length. Once completed, China's energy security, geopolitics and the security situation in the west region will get improved fundamentally. China–Pakistan Railway is Pakistan's national long-term strategic project, the Chinese government has also provided funds to Pakistan for upgrading its railway system, the railway geological survey and design for this railway has paved the way, and it is relatively mature to start planning and implementation.

For railways in Africa, China has traditional and realistic foundations with Africa in international affairs, economic aids and railway construction, although far from Africa; there is huge demand space for the construction of modern railways in Africa, and highly complementary with the railway industry of China, therefore cooperation on railway between China and Africa in the framework of China–Africa strategic cooperation has great prospects. With Premier Li Keqiang's visit to Africa, railway construction will again lead the process for China and Africa to further deepen the cooperation in all aspects. The official signature of the cooperation agreement by the Chinese and Kenyan governments on the Mombasa–Nairobi Railway marked the appearance of a new strategic opportunity for CHSR to go global, and we should seize the opportunity and press ahead quickly.

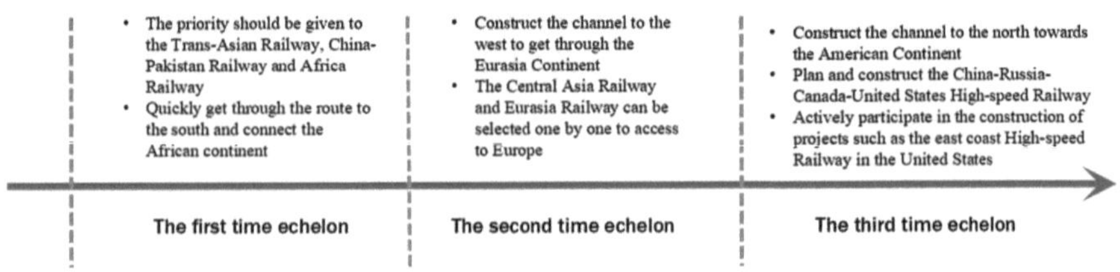

Fig. 5.5 Strategic arrangement for CHSR to go global: time echelon

The second time echelon: construct the channel to the west to get through the Eurasia Continent. The Central Asia Railway and Eurasia Railway can be selected one by one to access to Europe. The westward strategy is an important component of the national strategy, the integrated economic development of Eurasia Continent and getting through the railway channels to Europe are of important significance to the national defense and security, the unity of ethnic groups, the construction of the new Silk Road and the implementation of the strategy to develop the west. With the subtle changes in US–Russian relations, it is possible to promote railway cooperation within the framework of the SCO. At the same time, the operation of the China–Europe regular trains has further improved the atmosphere of cooperation. But it's also worth attention that, Central Asia and West Asia has always been a hotly contested place by the military forces, big powers are contesting severely in Central Asia, with constant conflicts in the area, therefore access to Europe via Central Asia will face a great challenge.

The third time echelon: construct the channel to the north to aim at the American Continent. This includes the China–Russia–Canada–United States High-Speed Railway under planning, and active participation in the construction of the High-Speed Railway along the east coast of the United States. The China–Russia–Canada–United States High-Speed Railway will link Asia with the North American Continent, and it will take only two days to travel from China to the United States.

In addition, we will march into the HSR market in North America, to compete for the east coast High-Speed Railway project in the United States. This project has just started, the US government will invest $117 billion in the coming 30 years, so the market potential is huge. Domestic enterprises including CRRC (formerly CSR) already made contact with it in the initial phase. However, the railway will run through Siberia and cross the Pacific Ocean, with a long line, many stations and complex terrain, the construction works will be arduous, and it will cover a number of countries, with different standards, therefore there will be considerable construction challenges, requiring planning for difficult situations and arrangement in advance. The CRRC (formerly CNR) has already formed a rolling stock company in the United States, and established the "Welding Structure R&D Center of CRRC-University of Michigan".

5.2.4 Strategic Routes

Route 1: "Products + Vehicle" Export.

This will start on the traditional markets in the Southeast Asia and South Asia, extend to the Central Asia, Middle East, Africa and South America, and then to the high-end markets in North America and Europe, to give full play to the comparative advantage and empirical curve effect and capability of CHSR equipment, and realize progress from export of single products to export of high end whole trains and locomotives. The types of export products include electric locomotive, internal combustion locomotive, motor train units, internal combustion train units, metro, light rail, passenger cars, freight wagons, engineering maintenance vehicle and other main equipment products.

Route 2: "Manufacturing + Service" Export.

It will change from trade of goods to the combination of goods trade to service trade, and extend from selling products alone to "products + service", i.e. from the export of mass transit equipment to export of railway systems. Focusing on creating value for customers, taking initiative to go into the user value chain, and providing "private tailor-making" according to the personalized demand of users, and solutions of mass transit equipment and system at high end and with high added-value and full-cycle one-stop service, will increasingly become the core competing power for the CHSR to go global. This will also lay a solid foundation for continually obtaining international orders.

Take MTU as an example, maintenance service has also gone global with the MTU. After MTU is put into operation, not only daily, weekly and monthly inspections are required, overhaul should also be performed. Product manufacturers know the products best, hence with the advantages in maintenance service. We can try to establish the 4S shops, so that customers can enjoy a series of supporting services such as maintenance, technical consultation, staff training, labor export and spare parts support.

Route 3: Export of Standards.

We will fully participate in the construction, financing and operation management for overseas railways, from export of "manufacturing + service" based on quality and customized products with high performance-price ratio and service advantages, to the export of standards based on technological advantages; and from building a station and a section of railway to final export of whole project, i.e. "going global" in all aspects of technology, capital, standards and talents.

Route 4: Establishing Internationalized Overseas Bases.

We will strengthen the construction of HSR production-learning-research bases from construction of domestic bases to cooperative cooperation of overseas bases, to support the development of overseas industrial cooperation, establish internationalized overseas bases, and realize the transition and upgrading from the export of products alone at first to export of products + services + technologies + base construction and project headquarters. With overseas bases, we will seek for new growth, obtain knowledge of cooperation partners, share development cost, disperse systematic risks on single market, avoid cyclic changes on single market, and strive to become first choice cooperation partner to push the development of local mass transit industry cluster. Further, the export of HSR will promote the internationalization of large equipment industry including the ordinary railway, inter-city, urban rail, metro, suspended rail and heavy-load mass transit equipment, and even the national communication, aviation and power equipment, to finally realize the strategic conception of high-end equipment of China enjoying world reputation.

Route 5: Owner + M&A + Strategic Alliance.

When selling products, technologies, services and base construction to all parts of the world, looking for excellent subjects related to owner for overseas M&A should also be part of the going global strategy of CHSR. It will be easy to obtain information, consultancy, skill and experience, and reduce various expenses in market transaction through M&A and strategic alliance, so that elements can flow bi-directionally or multi-directionally, to realize mutual complement of advantages, resources and risks sharing and organized advantages. Siemens of Germany, Bombardier of Canada, Alstom and GE of the United States, all these global transnational companies with worldwide influence have a history of M&A and alliance in their growth.

M&A subjects and alliance partners shall have fairly strong market competing power in R&D and manufacturing capability (such as design and manufacturing capability of IGBT), production and marketing and staffing, and can produce collaborative effects. Transnational M&A and strategic alliance are important components in the internationalization strategy of HSR enterprises of China, by selecting excellent alliance members through merging excellent subjects, they will continually allocate assets worldwide rationally, to promote industrial upgrading and raising the competing power on overseas market.

5.2.5 "Accessing", "Entering", "Improving" and "Approaching"

To enable the CHSR equipment to go global, we should adopt different strategies of "accessing", "entering", "improving" and "approaching" according to the access barrier of the target market, and the competing power and comparative advantage of own.

First, for the Market in Europe and the United States: "accessing".

For regions with average competition power but high barrier, such as Europe and the United States, the form of "accessing" will be mainly adopted. That is, a variety of flexible ways, such as trade, operation permit, outsourcing, buyer's credit, joint venture cooperation, capital operation, merger and reorganization, and strategic alliance can be adopted, to obtain technology, brand, sales channels or qualifications, to effectively evade trade and technical barriers, and go deeply step by step from points to areas, enter the market with products, to strive to break into the markets in Europe and North America with "hard power".

We should track main High-Speed rail projects, such as the High-Speed rail in California of the United States, HS2 in the UK, and High-Speed rail in Turkey. It is reported that the HS2 in the UK is divided into two phases according to plan: the first phase of construction will be 190 km of High-Speed rail from London to Birmingham, and the second phase will include two branches, respectively to Manchester and Leeds, with a total length of 568 km.

Second, for the market in Oceania: “entering”.

For regions with fairly strong competition power but moderate barrier, such as Oceania, the form of “entering” will be mainly adopted. That is, we will gradually expand the market by building R&D and production bases and in the form of joint venture, and enter it by financing, joint venture and M&A. We should give play to the overall advantages, strengthen and unify brand, and make overall planning for export trade, maintenance service and overseas investment for joint development, to quickly form complete overseas marketing and service networks, and consolidate our market in Oceania with soft power.

The main point of attention should be on the MRL signal system project in west Australia and Sperry detection vehicle project.

Third, for the market in Southeast Asia: “improving”.

For regions with fairly strong competition power but low barrier, such as Southeast Asia, the form of “improving” will be mainly adopted. That is, we will become localized enterprises, fully integrate ourselves into the local cultural, social and economic environment, give full play to the strategic advantage of value chain decomposing and integration, operation advantage formed by core knowledge and capability, the new type advantage of internalization, the system advantage of integrating supply chains, and the efficiency advantage formed by scale economy, range economy and speed economy, in the ways of international big enterprises and high-end brands.

We will actively carry out localized operations for overseas business, and establish economic, safe and stable overseas product manufacturing and supply bases. The “smart power” should be used to deeply develop the Southeast Asian market, to foster the ability of system total contracting, develop in the direction of integration of technology research and development, manufacturing and service, so as to gradually become a comprehensive solution provider for mass transit.

The focus of attention will be projects of Malaysia–Singapore High-Speed Railway, Indonesia High-Speed Railway, Nong Khai express railway in Thailand and China–Laos High-Speed Railway.

Fourth, potential market (Africa, South Asia, Latin America, Russia and CIS): “approaching”.

For regions with fairly strong competition power and high political (situation) barrier, such as Africa, South Asia, Latin America, Russia and CIS, the form of “approaching” will be mainly adopted. Africa, South Asia, Latin America, Russia and the CIS are potential markets for the HSR equipment export from China, and we should closely track the local political situation and movements, get familiar with the policies and regulations, taxation and other relevant information. For projects with mature conditions, we should cooperate with the country’s geopolitical and diplomatic strategy, and make more resource input, to meet the different needs of customers through merger and reorganization, strategic cooperation, outsourcing and so on, achieve breakthrough of all kinds of barriers with “discriminate power”, to cultivate these potential markets with great depth and breadth.

Main attention should be paid to the bidding projects of KOCHI shielding door and 152 sets of electric system for ICF, 40 traction systems for ICF of Calcutta metro in India, tamper wagon in Russia, and wireless signal system LCMS of South Africa TFR Company, etc.

In addition to the above four categories of markets, for regions with strong competition power and high security barrier (such as Central Asia and West Asia), we should make long-term follow-up and proper reservation, to access them prudently.

5.3 Strategic Main Points for CHSR to Go Global

CHSR going global is large system engineering at a high level and involving wide range and many key elements, and with far-reaching strategic significance, we must strengthen overall planning, and grasp the strategic points to push it ahead in an orderly manner. The following seven points are of vital importance.

5.3.1 Establish and Improve Authoritative and Efficient Coordination Mechanism and Management Service Platforms

CHSR going global is not only the development strategy of a specific industry, but also a national strategy and national will. To this end, we must establish and improve an authoritative and efficient coordination mechanism and management service platforms, to enhance top-level design, system layout and overall coordination, so as to avoid each department acting on its own and dispersion of resources and solve problems such as no macroscopic management institution at the national level, no systematic organization and specific functions with ministries and commissions, lacking in effective integrated management in the overseas financing by domestic financial institutions, and lacking in effective coordination between enterprises.

At the international level, we should actively build a multi-level policy exchange mechanism between governments. Governments of relative countries should jointly formulate plans and measures at multiple levels to promote regional cooperation, strengthen communication and coordination between the governments and cooperation with relevant international and regional organizations, improve the cooperation mechanisms with relevant countries on investment protection, finance, taxation, customs, personnel exchanges and other aspects, conduct timely consultation to resolve problems arising from the cooperation, to provide policy support jointly for pragmatic cooperation and implementation of large projects, provide comprehensive support and guarantee for enterprises to go global, and create a favorable policy environment.

To promote the CHSR to go global, China has established the inter-government mechanism of dialogue and cooperation with many countries, including the China–Europe Railway Working Group, the Railway Working Group under the Framework of China–US Strategic and Economic Dialogue, the Railway Working Group of Transport Cooperation Sub-committee of China–Russia Premier Regular Meeting Committee, the Greater Mekong Region Railway Union, and the Railway Working Group under the Central Asia Regional Economic Cooperation Mechanism. We should further improve the working mechanism of these institutions, include the key projects into the bilateral or multilateral agreements, and strive to obtain more support from the governments of relevant countries on market access, labor regulation, environmental protection, project contracting, and intellectual property rights, etc. We should accelerate the pace to consult, sign and implement with investment countries on agreement on avoidance of double taxation, to ease the burden on enterprises, make clear significant matters related to asset security and preferential policies, and improve the investment environment of enterprises in the countries involved. We should make active use of existing bilateral and multilateral cooperation mechanisms, to achieve effective communication and enhance mutual understanding, promote cooperation on production capacity in the field of High-Speed rail and promote sound development of regional cooperation.

At the national level, it is suggested to organize a full-time institution to guide the work for the HSR to go global, take charge of the top-level design for the internationalization of CHSR, overall deployment and medium and long-term planning for CHSR to go global, erect the foreign cooperation platform for the government and enterprises, and coordinate resources of all aspects with overall planning. In the meantime, we should complete the government administration and coordination mechanism and inter-ministry consultation mechanism, and make clear the specific duties and responsibilities and division of work of various governmental mechanisms such as the Ministry of Foreign Affairs, Ministry of National Defense, National Development and Reform Commission, Ministry of Education, Ministry of Science and Technology, Ministry of Transport, Ministry of Commerce and Ministry of Culture, the Railways Corporation and financial institutions such as China Development Bank in matters concerning "going global". We should give full play to the positive guiding role of traditional media and new media, strengthen public propaganda and safeguard the overseas image of Chinese railways. At the same time, railway counselors or commissioners should be staffed in embassies in countries with considerable railway projects.

At the enterprise level, under the existing railway system in China, the China Railways Corporation (hereinafter referred to as "CRC") should take the lead to establish coordination mechanism between enterprises, to shape the pattern for all participating parties to discuss matters jointly and share benefits and risks. In view of the development history of railways and HSR of China and the present reality, the CRC can surely undertake this mission. By leading the HSR enterprises, the CRC can better serve the overall interests of the state, and unify the state interests with interests of enterprises; it can give better play to the integration advantages of the whole industry chain, and better provide systematic overall solutions; it can better demonstrate its advantages in organization and management, and better agglomerate the overall resultant force in work, and it can push key planning, design and consultancy entities to go global, to realize development of foreign business in high-end fields. As the CRC has the advantages in organization resources in the industry, in domestic market and in comprehensive technologies, and also based on its historical relations with other HSR enterprises, it can perform energetic internal integration for consortiums of HSR enterprises, raise the work efficiency and lower the negotiation cost, which is also a unique advantage of China in the intense international competition.

The CRC will take the lead to establish and gradually complete the public service platform for the railway industry to go global. We can learn from the pattern of the public service platform of the Ministry of Commerce to go global, and comprehensively integrate information resources such as the government, chamber of commerce, enterprises, financial institutions and intermediary service agencies. Relying on the platform, we can do research and tracking for key overseas projects, to provide consultancy services in multiple fields of policy, law, market, access, and technology based on the information service and re-processing for enterprises to go global.

5.3.2 Build an International Cutting-Edge HSR Industry System and a Continual Technology Innovation System

Globally, a new round of innovation surge is coming, and the competition between countries has changed from that in resources field to that in industrial system and continual technology innovation. It was made clear long ago in the 12th Five-Year Plan of the state and the *Decision of the State Council on Speeding up Fostering and Development of Strategic Emerging Industries* that, we should energetically develop rail transport industry and further increase our efforts in indigenous innovation.

For the CHSR to go global, it is necessary to study deeply the industrial links of planning and consulting, investment and financing, design and construction, R&D and manufacture, operation and maintenance, education and training, system integration and project management on the HSR industrial chain, including environment evaluation and risk and efficiency evaluation, make clear the relationship between them, complete the management integration, technology integration and industrial chain integration, build an international cutting-edge High-Speed industrial system, High-Speed rail industrial chain and innovation chain oriented to the international market, and develop new advantages in industry international competition.

Guided by the innovation drive strategy of the state, and by formulating strategic plans for the development of major HSR equipment and policies and measures to guide the continual innovation and strategic development of industrial technologies, we will increase R&D input, raise the technological innovation level and the efficiency to convert innovation results, and enhance the motive force for sustained development and profitability, to create advantages for long-term competition. In the meantime, we will promote the innovative integration of technical standards with scientific and technological R&D, ensure protection of intellectual property rights, and raise the internationalization level of Chinese technical standards and the ability of technological adaptation to the conditions of host countries.

Through the comprehensive comparison of cutting-edge technology in HSR field in global range and the in-depth analysis of international HSR benchmark enterprises, we will continually deepen the indigenous innovation ability on key technologies and equipment, gain profound insights into development direction of technologies such as systems integration, car body design, bogie, traction and braking, train network control, key materials, aerodynamics, and operation and maintenance, foster core technologies, cutting-edge technologies and subversive technologies that will lead the future industrial development of HSR, and speed up the development of the technical standard system for CHSR based on our own technologies.

5.3.3 Promote Internationalization of Chinese HSR Standards, and Build Brands of CHSR

The process of CHSR going global will be accompanied by the export of Chinese standards and brands. In the domestic HSR construction, China has basically established a fairly complete standard system for the design, construction, equipment and safety of High-Speed railways and brands of railway equipment. We should further strengthen the development of technical standard system for the railway industry, and properly do the basic work conducive to the internationalization of Chinese standards. Also, we should energetically implement the standard internationalization strategy, brand strategy and intellectual property right strategy, and raise these strategies to the level of overall development strategy of CHSR.

We should compare and analyze the difference and equivalence with the advanced international standards, and promote the Chinese standards to go international. We should rationally demarcate the responsibilities of the State Railway Administration, China Railway Corporation and relevant manufacturing enterprises on formulating technical standards, further develop the own standards and codes of China and speed up the revision of important standards for the demand from the railway transportation and construction in China and the HSR enterprises participating in international competition according to the practice of international standard system and in conjunction with the actual global conditions. We will prepare the international versions of HSR standards and codes of China, promote the internationalization of standards of different spectrums, and promote the connection of HSR technical standards of China, to build up new competition advantages of enterprises with technological innovation and internationalization of standards.

The International Union of Railways (UIC) and Railway Cooperation Organization (RCO) are two main international organizations for railways, both are responsible for the formulation of railway standards. Within the scope of RCO, the work of High-Speed railway planning and standard formulation is still in the starting phase, so China should take an active part in this job, and strive to play a dominating role.

Also, attention should be paid to establishing and completing regional International High-Speed Railway Union, to participating in the establishment of regional railway standards, so that Chinese standards can first become international standards within the scope of RCO strategically. As a member of UIC and RCO, China should strengthen the communication and exchange with these two organizations and other institutions such as the International Standard Organization (ISO/TC269), International Electrotechnical Commission (IEC/TC9), European Standardization Organization (IEC/TC9) and European Standardization Organization (CEN/CENELEC), take an active part in the international standardization activities, and enhance the right of speech of Chinese railway in the formulation of international standards.

We should actively adapt to the requirements of foreign railway technology standards, and continuously improve the compatibility and extensibility of the Chinese technical standards on HSR. As is known to all, the early High-Speed rail equipment and technologies of China mainly came from Bombardier of Canada, Kawasaki Heavy Industries of Japan, Siemens of Germany and Alstom of France. With such background, the CHSR is well compatible with advanced technologies in the world in the main areas of project construction, MTU, train control and traction power supply, and this has laid a very good foundation for the internationalization of Chinese standards.

We will vigorously promote the HSR standard system of China and break through the overseas standard barriers through public relations work. We will, backed up by international projects, promote the use of Chinese technical standards on HSR, and in the process of participating in the construction of High-Speed railways outside China, we will have these countries adopt Chinese technical standards as much as possible. Also, opportunities of international exchange and talks can be used to publicize the Chinese standards. For example, in training professional technical personnel for other countries, we should enable them to fully understand and advocate Chinese standards.

We should firmly build up the awareness of brand. Full use should be made of the favorable factor that CHSR accounts for over 60% of total HSR mileage in the world and the comprehensive competence of advanced and complete technologies, good safety, reliability and compatibility, and high quality and performance-to-price ratio, to build up the brand of CHSR while promoting the internationalization of Chinese HSR standards, to raise the international publicity and brand influence of the HSR industry of China. In project construction and equipment manufacturing, we will brand the products, make the brands famous and internationalize the famous brands, to continually enrich and raise the technological and cultural connotation of HSR brands, and erect the high-end brand image of CHSR globally.

5.3.4 Speed up the Fostering and Training of Internationalized Talents for HSR

CHSR going global faces not only the powerful competition from HSR of Germany, France, Japan and Canada, but also the awkward situation of serious shortage in its own internationalized HSR talents. In fact, internationalized talents cannot meet the need in rapid expansion of overseas operations in both quantity and quality, becoming a bottleneck that restricts the further development of international operations. It is therefore urgent to speed up the fostering and training of high quality international talents for HSR, not only for coping with the realistic challenges and raising the comprehensive competing power of CHSR, but also for the sustainable development of the going global strategy.

First, build a multi-level system for fostering internationalized talents. The internationalized fostering plan for HSR talents includes the selection of people to be fostered, determination of training contents and research on fostering pattern. The role of professional railway colleges and schools, dragon' head railway enterprises and railway industrial associations should be brought into full play, and high quality resources of schools and enterprises will be integrated with the deep cooperation of professional colleges and schools and backbone enterprises as the backup, to build some national demonstrative (international) railway education and training centers, and reserve talents to support the smooth implementation of the going global strategy. At present, extraordinary actions should be taken to train two types of talents in urgent need, one is those for expanding overseas market who understand the politics, economy and humanistic conditions in target countries or regions, and are familiar with international economic operation rules and drafting of overseas bids and plans and bidding operation processes; and the other is those for overseas project management who are "trustworthy in work, proficient in business and language, and good at management and operation".

Second, innovate operation mechanism. We should adopt multiple ways to foster internationalized application-oriented talents, take initiative to connect with international standards for HSR, introduce advanced tutors and course systems from countries such as Germany and Japan, improve the quality of engaged expatriate teachers, and implement localized reform of some courses according to actual conditions. We will actively promote international and regional mutual certification of professional qualification for railway technical personnel of China. We will actively send students to main countries for production capacity cooperation by strengthening talent exchange and cooperation with international institutions such as UIC, and encourage excellent talents of China to go international. We will use digital, network and informationized means to carry out international remote teaching and MOOC learning.

Third, carry out international project exchanges and cooperation. Railway colleges and schools should actively join in and dominate international institutions such as "International Union of Eurasia Communication Higher Schools" and "Alliance for Rail Transit Application Technical Talents", promote the joint training programs with rail transit schools in countries along the lines, and build railway talent fostering and practice bases through the joint construction of international cooperation schooling projects such as "East Africa (Kenya) Rail Transit High Skill Personnel Fostering Base" and "ASEAN Rail Transit Personnel Training Center". In the meantime, we should widen the vision of teachers, accumulate internationalized training experience, build teacher exchange platform, improve the target orientation and agreement in fostering internationalized personnel, and further know the demand from overseas rail transit market.

Fourth, pay attention to foster local HSR talents. HSR is a high technology and high intelligence intensified industry, and generally countries along the Belt and Road don't have sufficient talent to support the HSR industrial chain. Therefore, it is necessary to specifically foster and train personnel of various levels and categories from technology to management and from grass-root level to high level for host countries, with the emphasis on the front-line professionals. The training can be conducted either in China or locally, and in various ways, the method to train overseas students can be adopted, and united schooling can also be used, to implement academic education for students majoring in railway in host countries.

Fifth, consummate policies and measures to attract talents. We should further widen the channels to admit talents, simplify recruitment processes, and improve various incentive policies and associated service to attract excellent talents both at home and overseas, to engage a batch of high-end internationalized talents with international operation ability, familiar with international operation patterns and proficient in languages and culture of main countries. We will also employ more excellent management personnel and competent front-line staff in the host countries by means of transnational M&A and strategic alliance.

5.3.5 Enhance the Risk Prevention and Control Systems

The American Express West Company issued a public notice on its website on the evening of June 9, 2016, officially ending all activities with China Railway International (USA) Co., Ltd. to form a joint venture company to build the High-Speed passenger railway in the United States. It explained that, "the US Federal Government specifies that a High-Speed railway must be built by the United States, there are many difficulties to implement the project, and also, it will be very difficult for China Railway International to obtain relevant permit from the (US) government". China Railway Corporation later issued a statement saying that it would deal with the matter in accordance with the law. In fact, this incident reflected some common problems in the going global of CHSR, and they can be summarized as "insufficient risk evaluation".[3]

[3]Wang Mengshu: *Risk Evaluation should be Strengthened for High-Speed Railway to Go Global, Global Times* June 13, 2016.

HSR construction is highly risky, especially in other countries. Take the political risks in host countries as an example, the unknowable political situation in some countries can expose the HSR construction to potential political risk, and political upheaval, especially change of government in an abnormal way often leads to government policy adjustments. Possible interventions by big powers in or out of the region can make the complicated situation even worse. Most countries along the Belt and Road are developing countries, not quite stable politically, and the authority of the central government is relatively weak, tribal politics are prevailing, if the state regime changes, the project may have to stop.

In fact, the biggest risk for CHSR to go global now is political and geopolitical risks. In recent years, China's strong rise has made some countries particularly vigilant, some countries along the lines have expressed their doubt on the military intentions and geopolitical intentions behind China's development strategy, and even developed hostility, leading to sharp increase of geopolitical risks. Perhaps "economic security" is the best excuse for "political intervention" to the construction of HSR in their countries by China. In their cooperation with China for HSR, the consecutive "changes" in India, Vietnam and Thailand are clear evidences. The contradicting action of the Mexican government on China's High-Speed rail project was a reflection of the political struggle in the country, and fundamentally it was due to geopolitical causes.

In the process of "going global", CHSR must strengthen the risk prevention and control system in order to guard against the many risks represented by politics and geopolitics. First of all, we should make proper environmental research, feasibility study and risk assessment in the early stage of project cooperation, and determine the return mechanism. We should review the risk factors from political, geopolitical, economic, financial, social, technical, legal (including patent and property) and environmental aspects, promptly make overall evaluation of the potential risks for their prevention, and focus on the different risk factors at different stages of the project. On this basis, serious and scientific risk dispersing mechanism should be established, and necessary preventive measures should be taken in advance for the factors that may affect and hinder the implementation of the project.

Taking the prevention and control of legal risks as an example, we should thoroughly investigate and know the legal environment and operating environment of the host country. With the joint efforts of internal and external lawyers, the foreign capital access system, investment approval procedures, company setup procedures, company legal framework, tax system, labor and law system, environmental protection requirements, foreign exchange control requirements and other laws and regulations closely related to the business and revenue should be investigated, studied and confirmed in detail. We should establish and improve the mechanism for dispute settlement, reduce the cost of dispute settlement, and use international rules to safeguard lawful rights and interests.

Second, strengthen the internal risk prevention and control mechanism in enterprises. We should pay attention to risk assessment and management, strengthen the duties and responsibilities of internal audit department and risk control committee. We should establish a powerful financial control system, establish an efficient and comprehensive budget management system, and manage all links in the operation in a timely manner through budget management. We should actively place overseas investment insurance, and establish a remedy mechanism against sudden occurrence of risks. We should also adopt local corporate governance strategy, strictly abide by the laws and regulations of the host country or region, and normalize the operation activities in overseas projects. We should pay attention to contract negotiation details and carefully review contracts. In addition, it is necessary to strengthen and improve the enterprise internal control environment, and develop the "criteria on business code of conduct", which should include: (1) the ethical policies, (2) the conflict of interest, (3) the policy to work as a director, (4) the policy on presenting gift and entertainment reception, (5) policy to participate in political activities, (6) multinational operation policy, (7) antitrust law, and (8) health policy, etc.

Third, establish a multi-level risk prevention and control service system of government, industry and enterprises. According to the principle of government guiding, industry coordination, enterprise acting as the main player, and prevention as the main, we should improve and complete the overseas risk prevention work plan, strengthen the multi-level organization and coordination function of the government and industrial associations, and build an intermediary service system that is marketized, socialized and internationalized for going global. We should give full play to the role of government at all levels and in all sectors, to improve the level of comprehensive services by government to prevent and control overseas risks. The Ministry of Commerce of China has been regularly issuing information such as *Guide for Outbound Investment by Countries, Risk Analysis Report by Countries* and *Investment Operation Barrier Report by Countries*, and this good practice is worth adhering.

5.3.6 Increase Financial and Banking Support

HSR requires huge initial investment, and lacking of funds and the immature investment and financing methods can seriously restrict its development. According to the estimation by the AIIB, the infrastructure investment, including HSR in Asia by 2020 is expected to reach $730 billion.[4] Although the Brics Bank, AIIB and Silk Road Fund can mitigate the investment and financing pressure to a certain extent, most countries along the Belt and Road, especially those neighboring with China, are developing countries, with limited economic strength. Also, the access of private sectors to the traffic infrastructure construction field is subjected to restriction to a certain extent in some of these countries, therefore HSR construction heavily depends on national finance.

At present, the HSR enterprises of China are facing significant capital shortage and financial risks, and it is imperative to solve problems of high financing cost

---

[4]Li Chenyang and Yang Xiangzhang: *Progress, Challenges and Prospects of Connectivity Construction of China with Surrounding Countries, Journal of Strategy and Decision-Making* No. 4, 2015, p. 3–16.

and high insurance cost. This requires the government to provide more banking and financial support.

First, the existing financial support policies need to be adjusted and optimized. The duties and responsibilities, resources and interests of the government, enterprises and financial institutions in different HSR projects should be defined, and be embodied in macroscopic policy, strategic planning, resources integration and regulatory examination. Inter-government HSR projects and railway projects in key and special countries (mainly developing countries) should be dominated and financed by the government, plus the policy of break even with meager profit. For strategic projects that cannot be covered by existing policies but really need support, special financial support policies should be studied and determined by overall planning on a case by case principle; for commercialized projects (mainly in developed countries), enterprises are the main players, therefore they should make decisions, bear profits and losses and risks by their own, and safeguard their rights by law. In principle, the government should support enterprises in the work of the project initial period by offering all capital or appropriate allowance for such expenses in all HSR projects.

Second, improve the banking and financial support policy system. The support by preferential loans should be increased to the going global HSR projects and major projects to promote international production capacity cooperation in railway, and preferential policies on financing quota and financing cost be provided directionally. A number of ways, such as helping financial institutions reduce financing cost, entrusting loan with foreign exchange reserves via franchised financial institutions, and overseas directional issuance of US dollar bonds with US government bond as guarantee, can provide relevant financial institutions more foreign exchange fund sources with low cost, to reduce the interest rate level of loans offered by Chinese financial institutions.

Bilateral investment protection agreement signed with foreign countries should be checked, and detailed rules set, so that Chinese enterprises making outbound investment can make good use of the bilateral protection agreement, and special bilateral agreement or coordination mechanism should be signed for key countries and areas. Overseas investment in Renminbi should be promoted, to help Chinese companies avoid the exchange rate risk in overseas investment. We will increase fiscal and tax support for the going global of railway enterprises, speed up the consultation and signature of agreement to avoid double taxation with relevant countries, and implement the income tax credit policy for overseas enterprises. The central finance will allocate special fund for special insurance, to improve the overseas railway investment insurance system.

Third, build a multi-party financial service system. We will jointly build a financial service system for going global with policy-based banks such as China Development Bank, Export–Import Bank of China, and China Export Credit Insurance Company and developing financial institutions as the main bodies, the commercial finance as support and extensively incorporating institutions such as commercial insurance companies, and other commercial non-banking financial institutions to participate it together.

At the national level, the quotas for sovereignty funds such as China Investment and policy-based funds such as Guoxin should be increased, to pay attention to HSR going global in the investment area, and the investment return expectation will be relaxed as appropriate. We will make good use of the large amount of foreign exchange reserve resources, encourage commercial financial institutions to carry out the syndicated loan, export credit and project financing, innovate financial products, and give full play to the role of financing and intelligence pooling. Support will be offered to enterprises having gone global to finance with the overseas assets and equities and mining rights as mortgage.

Fourth, guide financial institutions to provide all-round financial services. The government can coordinate to establish division of work and collaboration between the policy-based finance and commercial finance, and give full play to the advantages of "combination" of comprehensive financial business, to provide a package of comprehensive financial services covering financing, insurance, international settlement, financial information consulting for enterprises having gone global. Financial institutions should be encouraged to involve early in the development process of overseas railway projects, participate in specific pre-feasibility studies and feasibility studies, and help enterprises design feasible project financing structures.

Fifth, deepen the government and enterprise cooperation mechanism to share risks. Improve the capital replenishment and restraint mechanism of financial institutions and improve in all aspects the comprehensive strength and risk resistance of the financial industry; encourage Chinese Export Credit Insurance Company to design targeted insurance products; establish the risk compensation mechanism of financial institutions. Explore how to revitalize the fixed assets established overseas by the overseas cooperation zone and enterprises having gone global, so that they can become mortgaged property that can be guaranteed, and full use can be made of domestic foreign exchange reserves, to realize domestic loans with overseas guarantee and reduce the financing cost.

Sixth, support enterprises to "go out" to carry out overseas PPP projects. As the past single owner and government financing mode is hard to meet the needs of international infrastructure construction market, a series of problems, such as debt scale approaching the upper limit, government financing being restricted by regional policy, and low efficiency in infrastructure construction and operation management, has brought challenges to the pattern dominated by government financing. The development of project financing patterns such as PPP and franchising operation, and the diversification of financial investment main bodies is the inevitable trend of future development, and it is the inevitable choice for CHSR to enhance international competitiveness to adapt to market changes and innovative financing patterns.

5.3.7 From "Going Global" to "Entering Target Countries"

The criterion for judging the success of HSR going global is to really "enter" the target country. To smoothly go global, CHSR needs to build a positive and mutually beneficial international image. It must be committed to creating a favorable international political and economic environment and enhancing cross-cultural communication. To go global, CHSR certainly needs hard power and discriminate power in construction technology and equipment manufacturing, and should consider economic and commercial interests, but more important, we should realize that HSR is not only product export, but also cultural integration, therefore it needs more soft power and smart power, if we only focus on short-term benefits, it would just go to the opposite side.

For important node countries and target countries on Belt and Road, we should highlight the concept of "win-win cooperation, integration and connectivity". Most countries along the Belt and Road have the historical memory of Silk Road, to arouse the collective memory in these countries with the HSR construction and operation, will be the important contents of "people-to-people bond". It will be welcomed only when people-to-people bond has been established, otherwise it will be refused. How to "enter"? We need to define again the social cognition of CHSR going global in concept, expression and publicity, and replace the "invading and outputting" form with "integration and sharing" to win trust with good images.

HSR is a significant matter to any country, and a decision on HSR is essentially made by government. Once the decision is related to the government, the political system and political process of a country will inevitably affect the fate whether the HSR can enter it, so it is necessary to carefully deal with the relationship between the HSR and the government of the target country. Therefore, we need to fully listen to the positioning and appeal for HSR by the host country government, thoroughly analyze the basic national conditions of the host country, such as political, economic, social, religious, ethnic, population, land system, and historical and cultural heritage, because the demand for High-Speed rail of a country is closely related to the government will and the political environment, economic development and social structure of that country.

To enable the HSR to smoothly "enter" a target country, surely great importance must be attached to the social and civil actions. We should be able to properly use the forces of non-governmental organizations and the public. Establishing the partnership relations for joint development and win-win cooperation with stakes of local residents, communities and non-governmental organizations and obtaining trust and support from local public will probably be the key factor for CHSR to smoothly "enter" the target country. We should pay attention to collecting and mastering local economic and social information, fully consider the local political ecology, public opinion and attitude of all social strata, and listen to the voices of local industry association, the interest groups and academics. At the same time, we should give play to the role of overseas Chinese and strengthen the communication and exchange with overseas Chinese associations (including research institutions, regional organizations, chambers of commerce, clan associations, etc.).

In fact, Japan has done quite well in these aspects. In recent years, the Japanese government has vigorously promoted the "Shinkansen diplomacy", in the process of promoting railway infrastructure exports to Southeast Asia, it has paid attention to playing the role of "the Japan External Trade Organization" and "Japan Chamber of

Commerce" and other official and non-governmental organizations, and it also set up cultural institutions in Southeast Asian countries, to pave the way and erect bridges for Japan's rail transit equipment export. In the competition for Mumbai–Ahmadabad High-Speed rail project in India, the Japan International Cooperation Agency (JICA) did active lobbying in local administration, enterprises, schools and research agencies and departments, to push the relevant parties to actively accept the Japanese plans.

To smoothly enter the target countries, the CHSR enterprises to go global must give play to the comparative advantage of both parties, and implement the localizing development strategy according to the actual local conditions. Firstly, help the host countries to develop their industries. We should make full use of the host country's natural and social resources, investigate and study the progress and depth of the local industries, analyze the present situation of the target market related to the railway industry, cultivate and support upstream and downstream enterprises associated with the railway industry chain and the research institutions, financial institutions, industry associations in relation to their development, to promote the formation of the railway industry and related industries in host countries.

Secondly, we should identify local industrial fields and directions, clarify industry cultivation objectives and build value-added production chain. This includes the use of resources of the project countries, engaging local technical experts and consultants, setting up local subsidiaries or technical research and development centers, and building construction materials and structural members and parts factories. We should take initiative to undertake project management and its late period sustained development, strengthen the technology transfer and experience sharing while safeguarding the core interests of enterprises, improve the equipment supporting ability of local industries, and drive the local area to expand employment.

Furthermore, we should actively fulfill corporate social accountability, give full play to the advantages of enterprises to carry out public welfare activities, and vigorously develop the ability of enterprises to integrate into local society and contribute to the economic and social development of the host countries. We should strictly abide by the laws and regulations and regulatory rules of host countries, and attach importance to labor and environmental protection, to become respected enterprises in local areas.

5.4 Strategic Actions for CHSR to Go Global

In view of various matters faced by CHSR going global, the work in the following six aspects should be accelerated to strengthen relevant policy support at the national level.

5.4.1 Set up a National Level "CHSR Going Global" Leading Group

HSR going global is a strategic action that reflects the will of the state and the behavior of the government. It is a complex and huge system engineering project. Therefore, it is highly necessary to set up a national level "headquarters"—"CHSR Going Global" Leading Group. In actual operation, it is suggested that the existing railway "going global" work coordination group should be upgraded to the "CHSR Going Global" Leading Group, with the leader of the State Council as the Group Leader.

The mission of the leading group is, at the strategic height of "overall consideration of both international and domestic situation and comprehensive utilization of international and domestic resources", actively and effectively promoting the HSR to go global by relying on the national power, in the grand background of implementing the Belt and Road Initiative. It should coordinate various parties, make planning in advance, and push it forward in an orderly manner from multiple dimensions of politics, economy, industry, equipment, investment and financing, diplomacy, national defense and culture. It should strengthen dialogue and consultation among governments and establish a regular communication and coordination mechanism with the target countries.

The functions of the leading group: establish the strategic goal, route and measures for going global; establish and complete the national overall planning and coordination mechanism, strengthen policy support and complete the associated service system; coordinate the relevant ministries and commissions to effectively carry out "HSR diplomacy", and smoothly promote international production capacity cooperation on HSR; and strengthen the exchange and cooperation with international organizations such as the UN, World Bank, AIIB, and UIC.

A full-time office shall be set up under the leading group (tentatively in the National Development and Reform Commission, and the main work should be undertaken by the National Railway Administration), with sufficient funds to ensure its operation. The existing five management and coordination mechanisms for going global should be simplified, to implement matters and working responsibilities for overall planning and coordination and complete the mechanism for communication on specific matters and classified guidance by optimizing work procedures and holding periodical meetings. It should organize professional teams to deepen decision-making support and research; optimize the cross-sectoral international coordination mechanism, coordinate China Railways Corporation and related High-Speed railway enterprises, China Development Bank and other financial institutions, make overall planning to guide the CHSR to go global in a targeted, planned, step by step, and organized manner with emphasis.

5.4.2 Establish the Portal Think Tank for CHSR Going Global

For the HSR to go global, we should gather the research forces of the government, colleges and universities, research institutions and industrial sector, recruit and engage a group of full-time and part-time experts from HSR industry enterprises, financial institutions, legal affair institutions, national (regional) research institutes and other units, to establish a national level portal think tank for CHSR going global. By giving full play to the professional competence and comparative advantages of the members of the think tank, we should plan the top-level design and strategic deployment for CHSR going global. The think tank should focus on the three levels of mechanism, industry and enterprise to carry out research work.

At the mechanism level: for the HSR to go global, we should conduct in-depth researches on how to build the decision-making platform, integrate the implementation team, allocate administrative resources and re-build the governance mechanisms of business process flow; how to building the working mechanism with consistent target, powerful commanding, capable efficiency, balanced power and responsibilities and smooth process flow; and the mechanism for proper combination of the strategic leading principle, systematic planning principle, key point breakthrough principle, resources concentration principle, controllable risk principle and win-win cooperation principle for HSR to go global.

We should conduct in-depth researches on how to form the collaboration mechanism of government taking the lead, enterprises as the main body, financial support, market operation, multi-party joint development and industrial alliance; how to form the linking mechanism with the state taking the lead in coordination and division of work, the finance, manufacturing, construction and commerce and trade supporting each other, and relevant enterprises in the railway industry forming consortiums for different HSR projects; and conduct in-depth research on the economic and diplomatic policies, financial policies, taxation policies, banking and foreign exchange control policies to promote and push HSR to go global, especially targeted, breakthrough and operational preferential and associated policies.

At the industrial level: we should properly analyze demand on target market, classify markets in more detail and identify effective market. In-depth analysis should be made for the target market on the political and legal environment, economic and technological environment, social and cultural environment, natural geographical environment and market competition environment, to reveal the key areas, key projects and key direction for HSR to go global.

Strategically, we should further study how to maximize the overall superiority of the HSR industry of China and the HSR industry chain integration capability, improve capability of global allocation of resources, full industry chain management, key value chain control, core technology control and high value-added profitability, upgrade the internationalized comprehensive competing power in all aspects, to form the international level industrial power, unified our pace to gain resultant forces for transnational operation worldwide, so as to raise the overall strength and EPC capability for going global.

Tactically, we should further study how to consolidate the power of the industry with the domestic railway industry union as the basis, and build consortiums covering survey and design institutes, project construction enterprises and equipment manufacturing enterprises, to form the systematic capability integrating the project initiation, survey and design, investment and financing, project implementation, equipment supply, completion and acceptance, combined testing and commissioning, safety assessment and operation management.

In actual operation, we should further study the strategies and innovation and cooperation patterns in the equipment manufacturing sector and project construction sector with different measures, and put forth suggestions on practical operation for the following issues: how the equipment production capacity export can combine the globalized operation with localization work, to push forward the coordinated progress of multiple industrial sectors; how to give full play to the leading and piloting role of consultancy and design as the soul in construction of HSR projects with design and consultancy taking the lead and the construction works as backup in project construction, to realize the advantage complement in resources integration for projects.

At the enterprise level: we should study the strategies and overall solutions for enterprises to enter the target countries, with emphasis on revealing various potential risks in going global and the strategies to cope with, avoid, manage and control risks. We should prepare research reports for railway to go global by countries in the principle of conducting research and developing specific strategies for each country, and work out on this basis the targeted technological integration plans and measures on negotiations, public relations, commerce and legal affairs for entering, so as to ensure mastering the conditions in the target countries and apply strategies precisely.

Special attention should be paid to the issue of enterprise ability construction, so that efforts can be made to improve the global control power, market expansion power, resource allocation capability, risk prevention and control power, value chain competitiveness and brand shaping capability of enterprises, and to upgrade the capability of enterprises on globalized operation, local business operation, digital management, intelligent manufacturing and diversified development.

5.4.3 Organize International Industrial Alliance for CHSR Going Global

Industry convergence, industrial cluster and market competition are important factors for the success of HSR going global. To reduce the excessive and bad competition between domestic enterprises, coordinate internal enterprises to go to international market with concerted efforts and joint forces, and participate in overseas competition in groups, the international industrial alliance for CHSR going global should be established as early as possible, to form an organizational collaboration system covering the full life cycle of HSR industry, and realize linked operation of the industry and export with interaction. With this alliance as the basis, an alliance of permanent members of "CHSR going global forum" should be set up, to assist the in-depth advancement of the national strategy in a normalized manner.

5.4.4 Create International Talent Education Basis for Rail Transport

To go global, talents are the primary resources and the most powerful support. To enter target countries, the fostering and training of talents is conducive to the cultivation and development of overseas markets, and to the acceptance by target countries before products entering them. For the CHSR to go global, we should first foster and train talents, and efforts should be made to train both domestic HSR talents for internationalization and HSR talents in target countries, the education atmosphere to share the HSR technological results should be actively cultivated, to build the talent fostering and training system to support the development of world rail transport and form a "pool" of talents for sustained going global.

Therefore, it is suggested to start the construction of international talent education base for Belt and Road rail transport. This base will consist of five centers: international education center, rail transport exhibition center, rail transport experience center, intercultural exchange center, and soft science research center.

We should focus on implementing a batch of education cooperation projects, cultural cooperation projects, study abroad and visiting plans, and science and technology cooperation plans, to give play to the role of education as bridges, to spread the Chinese culture, tell Chinese stories, transfer Chinese value, draw closer the psychological distance of all walks of life between China and countries along the lines, and foster personages knowing China, friendly to China and willing to help China. More support should be offered to underdeveloped countries and regions in fostering and training HSR talents, to provide scientific and technological and manpower support to them in developing HSR.

Advance the internationalization of HSR education curriculum. We should strengthen the interaction and communication in developing international HSR education curriculum and the teaching materials, and intensify our efforts in digital education to realize the joint construction and sharing of education resources. We should concentrate our efforts in the academic and non-academic education projects for rail transport talents in countries along the Belt and Road, develop the Chinese standards for international rail transport personnel training, and foster and reserve a batch of rail transport talents with strong theoretical basis and engineering practice ability for these countries.

Talent fostering and training is the basic, pioneering and forward-looking work for CHSR to go global, and it is also a strategic, urgent and challenging work.

Through extensive investigation and research and full justification, the "22441" talent pioneering strategy can be implemented to solve the talent bottleneck for the going global strategy. The connotation of "22441" is as follows:

"2 systems", i.e. building up the "fostering system" and "training system" for railway internationalization talents.
"2 certifications", i.e. the construction of "professional education certification" and "occupational certification" for internationalization of HSR. They include setting internationally essential equivalent professional education certification and occupational certification standards, systematic design of certification systems, organizing the implementation of the certification process, and determining and issuing certification conclusions, to provide scientific criteria for personnel fostering and training quality in High-Speed rail education.
"4-position collaboration", i.e. establish the collaborated fostering mechanism of the government, enterprises, schools and research institutes.
"4 aspects of overall planning", i.e. overall planning for the domestic and overseas, the current and long-term, the online and offline, and the general coverage and highlighting main points.
"1 alliance", i.e. establishing the international HSR education alliance, to build up the information communication and resources sharing platform, establish manufacturing and teaching integrated and collaborative innovation mechanism, promote international exchange and cooperation and foster and train internationalized HSR talents jointly.

5.4.5 Construct "Rail Transport National Laboratory"

General Secretary Xi Jinping has pointed out that it is important to accelerate the construction of an innovation base platform led by national laboratories, so that it can serve as an important carrier to seize the commanding heights of scientific and technological innovation. In order to upgrade the core competitiveness of CHSR, we should, guided by the national goals and strategic demands, aim at the four strategic directions of advanced rail transport, i.e. the rail transport system security technology, comprehensive system efficiency upgrading technology, system sustainability technology and system interoperability technology, lead a new paradigm of technology innovation in rail transport field, take the high ground of transportation system technology, standard, equipment and integration capability system, and accelerate the construction of "rail transport national laboratory", to provide powerful support to CHSR for its going global.

We should make use of the opportunities in the construction of ten major areas when the state energetically pushes forward "Made in China 2025" initiative, speed up the application of new materials, new technologies and new processes in advanced rail transit equipment, make breakthroughs mainly in systematic security guarantee, energy conservation and environmental protection, and digital intelligent network technology, develop advanced, reliable and applicable products and lightweight, modular and serialized products. We should research and develop a new generation of green, intelligent, High-Speed and heavy load rail transit equipment system, to provide customers with overall solutions for whole life cycle of system, and build the world's leading modern rail transit industry system.

At present and in a period to come, we will speed up the scientific research in cutting edge science and technology and key area such as air, vehicle and metro information integrated rail transport safety and control key technologies, rail transit system safety guarantee technologies, High-Speed passenger transport equipment and technologies for speed over 400 km/h, key technologies in maglev transportation system, extra-high speed vacuum pipeline maglev transportation and dynamic model wind tunnel, to realize technology breakthroughs and technological leap forward in HSR.

5.4.6 Establish "Overseas Investment Venture Fund for High-Speed Rail Projects"

We should explore safe and efficient ways of investing and financing to cope with the risks and challenges posed by unfamiliar environment for HSR going global. In view of the high investment risks of overseas High-Speed railway projects, the state should establish "overseas investment venture fund for High-Speed rail projects". At the same time, full use should be made of foreign exchange reserves, and the foreign exchange reserves should be well combined with the transnational operation strategy of HSR enterprises and the international operation strategy of financial institutions, to promote HSR enterprises and financial institutions to go global together.

We should strengthen the cooperation of banks and enterprises, continue to give play to the pilot and basic role of policy-oriented financial institutions, coordinate the national banks to provide preferential project loans (policies such as buyer's credit and seller's credit), offer more support to overseas High-Speed rail projects, and increase the financing quotas for overseas High-Speed rail projects. The China Development Bank and Export–Import Bank of China can guide and support the HSR to go global by supporting in priority the first obtained strategic projects in the Belt and Road strategy. We should research and build regional financial platform for HSR to go global, to specifically support overseas High-Speed rail investment projects, and guide and support enterprises to go global in the forms of equity, debt, industrial funds, buyout funds and M&A funds.

Relying on the Silk Road Fund and the AIIB, we will improve the "construction+investment and financing (EPC+F)" system of High-Speed rail. On the one hand, we will reduce the risk of foreign exchange reserves of China, accelerate the pace for RMB to become internationalized, and preempt the right to establish new rules for global trade and the right of control, pricing and resource allocation. At the same time, we will enhance the financing capacity on the international financial market. The direct financing on the developed securities market and banking sector abroad will be used, or advanced means of financing such as financial leasing of equipment and facilities, ABS (asset-backed securitization), leveraged buyouts, syndicated loans, and setting up international railway industry investment fund will be used, to promote the going global HSR projects to expand with diversified financing to the international capital market.

On the other hand, in conjunction with the investment and financing environment in the target countries in East Asia, South Asia, Central Asia and the Middle East, and according to the actual situation of owner countries, we should flexibly use investment and financing patterns such as BT, BOT, BOOT, and TOT (Transfer–Operate–Transfer), TBT (TOT+BOT, a financing pattern by combining TOT with BOT, and with BOT as the main), PPP and TOD. The BT pattern can be used for countries with fairly good financial capacity; for projects with large traffic flow and good business environment, the TOD pattern can be used, to realize the innovative return of business through land and property development.

In combination with patterns such as “HSR for resources” and “market for funds”, reduce the economic risks to a controllable range. We should study and popularize the pattern of “resources for HSR”, and strengthen the research on innovative cooperation pattern for resource-rich countries with backward economy. For example, exchange of petroleum and natural gas in Central Asia and Europe, minerals in Africa, potash minerals of Myanmar, rubber and rice of Thailand for HSR, and HSR for market and mutual political assistance can get economic or strategic resources for China’s development, to achieve win-win strategy.

Enterprises should also continuously optimize their own investment and financing structures such as their own capital, bank loans, capital market financing, government appropriation subsidies and private lending. In particular, most HSR enterprises in China have become transnational company, and they should carry out overall financial management planning, and take into general account their capital structure and financing arrangements.

Part III
Undertakings of Universities in the High-Speed Railway Era

The development road of China's High-Speed rail industry from the scratch, from weak to strong, and from following others to taking the lead, is a process of fostering the first-rate innovative talents for rail transport in China, and also a road of exploration to run research-based universities with distinctive features and powerful competence of rail transport. Transportation changes the life and education achieves the future. It is the responsibility, undertaking, and mission of institutions of higher learning to foster high-quality HSR talents, to push the integration, connectivity, and joint development of countries along the Belt and Road.

Chapter 6
Higher Engineering Education in the Waves of "Reindustrialization"

6.1 Engineering Science and Technology: The Power Motive Force for the Evolution of Human Civilization

6.1.1 Outstanding Engineering Technology Achievements in Ancient China

Engineering, science and technology has been a powerful driving force in the evolution of human civilization since man learned to use tools 2.5 million years ago. The Dujiangyan Irrigation Project built during the Warring States period has benefited the land of Sichuan for two thousand years, and the Great Wall, with construction starting from the Western Zhou Dynasty to the Ming Dynasty, "for a period of two thousand years and extending one hundred thousand li", has embodied the remarkable achievements of Chinese ancient construction and engineering technology.

Our ancestors also created remarkable outstanding achievements in various engineering fields such as ceramics, paper making, printing, smelting, casting, salt mining, flood control, irrigation, textile, printing and dyeing, machine building and construction. The famous British science historian Joseph Needham said in *Science and Civilization in China* that: "from the 1st century AD to the 18th century, at least 26 scientific and technological inventions spread to Europe and other places from China, they played an important promoting role in the development of world history."

6.1.2 The Industrial Revolutions Have Demonstrated: Powerful Industries Means a Strong Nation

In the middle of the 18th century, the invention of steam engine and textile machine triggered the first industrial revolution, turning the world from manual production to

F. Xu, *The Belt and Road*, https://doi.org/10.1007/978-981-13-1105-5_6

mechanical production, thus the mankind entered the industrial age. From the late 19th century to the first half of the 20th century, generators, internal combustion engines and chemical industry led to the second industrial revolution, ushering in the era of electrification, atomic energy and aviation and aerospace.

In the second half of the 20th century, the electronic computer, Internet and information technology caused the third industrial revolution, greatly improving the level of social productive forces, and pushing the mankind into the information and intelligent era. In recent years, marked by digitalized production, the fourth industrial revolution gave rise to a new round of industry transition, quantum information technology, space engineering technology, new biological technology, new energy and new materials technology constantly emerging, and a worldwide industrial structural revolution is afoot.

Reviewing the history of the evolution of human civilization, from the nomadic hunting or fishing to plantation, from the benighted barbarism to civilization, and from industrial civilization to the digitized information age today, we can see clearly that every significant change of human society, is always accompanied by the birth of some major technologies behind it, and those major technologies have become landmarks for us to know the history of the past.

From the rise of the UK in the first industrial revolution, to the transcendence by Germany and Japan in the second industrial revolution, and then to the third industrial revolution when the United States became the leader, until the ongoing fourth industrial revolution with powers rising side by side and in particular the meteoric rise of China, it has revealed profoundly that, the flourishing of manufacturing industry and strength is one of the decisive forces, and the simple truth "powerful industries means a strong nation" remains true throughout the history.

6.2 Big Nation Manufacturing Industry Strategy in the Post-globalization Era

Along with the socialization of production, trade liberalization, business electronization, operation virtualization and economic integration, the world has come to the era of globalization, and the most important feature of globalization is the prevailing virtual economy. Since the 1970s, the United States, Japan, Germany and other highly developed industrial countries implemented the policy of "de-industrialization", and realized the transformation and upgrading of their industrial structure through gradually transit of equipment manufacturing industry overseas.

The outflow of manufacturing industry resulted in serious "financialization" and hollowing out" of real economy in the United States and other developed countries, the virtual economy with the capital market as the main body gradually took the dominant position, the false prosperity of capital market and the shrinking of real economy eventually led to the outbreak of the global financial crisis.

The global outbreak of the financial crisis has given rise to a deep reflection on the virtual economy without the support by the real economy, thus triggering the wave of "reindustrialization". This wave is accompanied by the corresponding world geo-political reconstruction and rebuilding of international economic and financial order, indicating that the world is entering the initial period of a post-globalization era. After the international financial crisis in 2008, in order to solve the problem of massive unemployment caused by the economic crisis, the United States, Japan, Germany and other countries formulated corresponding policies to start the process of "industrialization", to carry out industrial transformation and upgrading, and direct the return of overseas manufacturing industry back home.

6.2.1 The "Re-industrialization" in the United States

After the financial crisis and the Internet bubble, the United States focused on readjusting the economic development strategy, and promptly launched the reindustrialization strategy to revitalize the American manufacturing industry. In 2009 and 2010, it issued the *Framework to Revive American Manufacturing Industry* and the *Act of American Manufacturing Industry Promotion*, in 2012 it put forth the *American Advanced Manufacturing Partnership Program*, and in 2013, it further launched the *National Network for Manufacturing Innovation*. The US manufacturing sector accounted for 15% of GDP in 2013, and the population of direct and indirect employment rose to 17.03 million, a staggering 49% increase from the 11.6 million at the outbreak of the financial crisis.

6.2.2 "Industry 4.0" of Germany

The manufacturing industry of Germany is powerful, ranking at the world top in quality and technical standard. In the face of the United States reviving its manufacturing industry and the rise of manufacturing powers such as China, to ensure the future for the German manufacturing and stabilize its leading position in the global manufacturing sector, *Germany issued the High Technology Strategy 2020 Action Plan* in 2010, launched the *Nanotechnology 2015 Action Plan* in 2011, put forth the *Proposal on Implementing the "Industrial 4.0" Strategy* in 2013, and formulated eight priority action plans. The core of the "Industry 4.0" of Germany is to build the Cyber-Physical Systems (CPS), which is aimed at creating new standards for intelligent manufacturing through smart factories, smart production and smart logistics.

6.2.3 The "New Industries" and "Future Industries" of France

In September 2013, France launched the "New Industrial France" strategy in the situation of the double continual declination in the industrial added value and the employment proportion resulted from the "de-industrialization". This ten-year strategy will mainly solve three big issues of energy, digital revolution and economic life, including 34 specific programs such as renewable energy, electric plane and a new generation of aircraft, smart grid, biological medicine technology, the Internet of things, robot, augmented reality technology, and non-contact service. In recent years, France has made a profound readjustment to the "re-industrialization" policy, and proposed the "new industrial France II" program and "future industry" program. Through these programs, France endeavors to rebuild its industrial power and keep itself at the top of the global industrial competitiveness echelon.

6.2.4 The "New Growth Strategy" and "New Robot Strategy" of Japan

In the 1980s, Japan created the world's most impressive achievements in household appliances and automobiles, and its quality revolution was unique and distinctive in the world manufacturing market. However, due to the excessive dependence on sophisticated techniques, and the ignorance of the new subversive technology innovation, especially due to the lack of market acuity, it missed the rare first opportunities for further development. With this painful experience, Japan released the *New Growth Strategy* in 2010, and the *New Robot Strategy* in 2015, to accelerate the development of strategic emerging industries represented by the robot industry.

6.2.5 Made in China 2025 Initiative

China's manufacturing industry has undergone profound changes in the past two decades and more, having made great progress in the leap forward journey from the "world factory" to the "high-end equipment provider" including High-Speed rail, and won the world surprising reputation in fields such High-Speed railway, long bridges, household appliances, mobile phones, satellites, unmanned aerial vehicle, supercomputation and hydropower. In the wake of the world financial crisis, China has further strengthened its resolution to transform from a "big manufacturing country" to a "manufacturing power" and further to a "smart manufacturing power" and "China creation".

In 2015 China issued the *Made in China 2025 initiative*, to implement key projects in manufacturing innovation center, intelligent manufacturing, strong industrial base, green manufacturing, and high-end equipment, and make clear the top ten key breakthrough areas including a new generation of information technology industry, high-end CNC machine tools and robots, aviation and aerospace equipment, marine engineering equipment and high technology vessels, advanced rail transportation equipment, energy saving and new energy vehicles, power equipment, agricultural machinery and equipment, new materials, biological medicine and new performance medical apparatus and instruments, as well as technology development roadmap for 2020 and 2025 (Fig. 6.1).

In general, in a new round of wave of "industrialization", the world manufacturing powers focus on reshaping their competitive advantages, vigorously develop key technology of robot, intellisense, intelligent control, micro-nano fabrication, complex manufacturing system, develop the key equipment and processes of major intelligent complete sets of equipment, optoelectronic manufacturing equipment, intelligent robot, additive manufacturing, and laser manufacturing, and advance the domestic manufacturing in the direction of networkization, digitalization, intelligentization and greenization, to promote innovation in the domestic manufacturing production mode and industrial form.

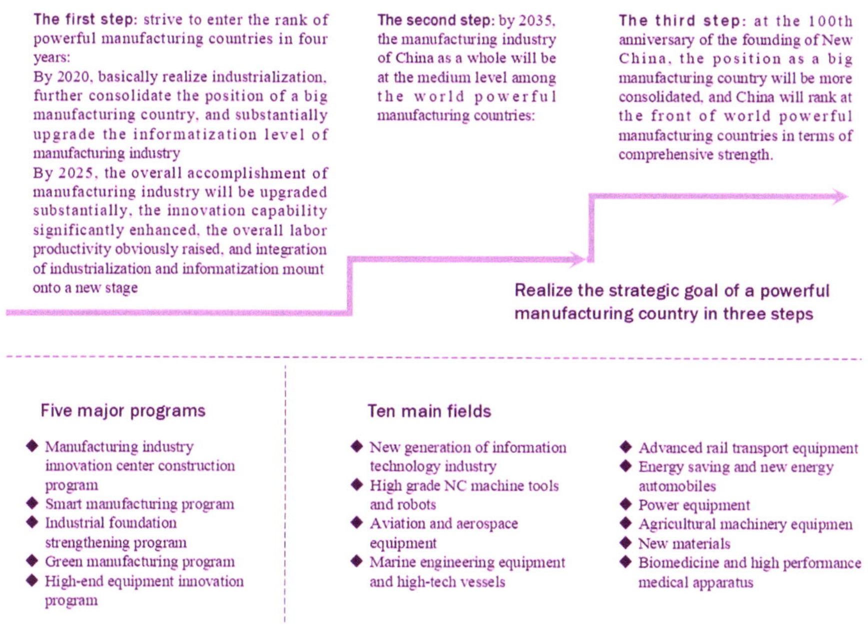

Fig. 6.1 Made in China 2025 initiative

6.3 Profound Transformation in Higher Engineering Education in World Important Economies

To adapt to the wave of "reindustrialization", important economies in the world, especially developed countries in Europe and America launched reform in tertiary education one after another. They build up general education training systems, enhance the engineering practice ability, improve the mechanism of collaborative education of enterprise, school and research, increase the engineering leadership, highlight the scientific and technological innovation consciousness, and strengthen the fostering of engineering and technical talents, to meet the demand for engineering and technical personnel in the new situation.[1] The international engineering education is mainly divided into American system and European system.

6.3.1 The United States: The Washington Accord

The tertiary engineering education in the United States in transformation energetically advocates cultivating engineering talents with capability in five aspects, namely, design and research, interdisciplinary learning experience, innovation and entrepreneurship, global vision and service learning.

In 2005 the National Academy of Engineering published *Training Engineers for 2020: Transform Engineering Education for the New Century*, putting forward the concept "engineering with a big E", requiring the students to pursue for engineering opportunities through general education, advocating "engineering degree + (humanities + social and management science, etc.) + enterprise joint training", and encouraging to cultivate the students' creative ability by developing "problem-centered" practicality courses. At the same time, it attaches great importance to the internship of engineering students, and takes the internship achievement as the core part for obtaining degrees, to guide the students to apply engineering technology and engineering tools to solve practical problems.

Massachusetts Institute of Technology published in 2014 the *Future of the MIT Education*, strongly advocating not only online learning, to set up the global community for learners, but also hands-on practice, experimental learning, blended learning, and experienced teaching, calling for pursuit of excellence. In 2015, the United States explicitly proposed to train at least 20,000 "big challenge engineers" in 122 engineering schools in the next ten years, to solve major engineering issues in the future.

In the reform and transformation of international engineering education, the Washington Accord (WA) played a very important role. The WA is an agreement concluded by six English-speaking countries including the United States in 1989 for mutual recognition of four-year bachelor degree higher engineering education, the parties to the agreement mutually acknowledge the accreditation policies, standards, proce-

[1] Fei Xu: *Run First-rate Engineering Education and Foster Excellent Engineering Talents, Research on Higher Engineering Education* No. 6, 2016, pp. 1–6.

dures and results on the professional engineering education schemes of other parties, thus building a global framework for higher engineering education accreditation. The accord is highly authoritative and influential internationally, and by now, it has 18 official member countries.

The WA has two outstanding characteristics: first, "take students as fundamental", emphasizing the standard "based on student learning achievements"; second, users participate in the authentication and evaluation, emphasizing the effective connection between industry and education. As a kind of innovation of education paradigm, the professional authentication is based on the three basic concepts "student-centered", "result oriented" and "continual improvement", a distinctive contrast from the traditional education of content-drive and attaching great importance to input. These two characteristics represent the direction of international engineering education transformation in the future.

6.3.2 Germany: Craftsmanship

Germany, with the tradition of attaching high importance to higher engineering education, emphasizes more under the new situation in the following three aspects: one is paying attention to the combination of science with technologies, the combination of basic research with applied research and development research, with focuses on the technologies and development research, so that students can better grasp the scientific research methods in the field of engineering technology, have skilled technical implementing ability at a high starting point while possessing solid theoretical foundations. The second is linking the department and discipline setup with the social demand, and maintain close contact between institutions of higher engineering education with business circle, to cultivate engineering and technical personnel according to the need of the national industrialization process and enterprise development. The third is paying attention to general education and students' learning of humanistic and ethical knowledge, and pay attention to the promotion and inheritance of "craftsmanship", and devoting to the cultivation of well-trained excellent engineers.

6.3.3 The United Kingdom: The "Sandwich" Education Pattern and Professional Certification System

The UK explores the implementation of a "sandwich" education pattern, in which students alternately fulfill work and learning tasks. First, students must complete the knowledge required for one year work practice in enterprises, followed by study for two to three years, and finally they will do the practical operation in enterprises for one year. In 2011, the UK started implementing the "Program to open and understand manufacturing industry" and invested 125 million pounds to build the "advanced manufacturing industry chain".

The implementation of professional authentication system is an important institutional arrangement of in the higher engineering education in the UK. The UK established a three-level quality authentication management system from top to bottom with Higher Education Quality Assurance Agency (QAA), the Engineering Council (ECUK) and Joint Board of Moderators (JBM) of various fields, to ensure the quality and standard in engineering education disciplines and degrees. The system of registered engineer is implemented in connection with the authentication system, including the registered chartered engineer (CEng), registered assistant engineer (IEng) and registered engineering technician (EngTech). In addition, the focus on the cultivation of precision manufacturing and digital creative talents is also a major feature of the transformation of higher engineering education in the UK.

6.3.4 France: Elite Engineers

Early in the 18th century, the Bourbon Dynasty of France built a batch of specialized schools, and from the Napoleon period until the present, France further strengthened their application functions, such as engineer schools of Paris Road and Bridge, Paris Mining School, Paris Superior Normal School, and Paris Polytechnique School, these schools, with admission ratio of only 10% and high student-faculty ratio (an average of 1:5), are committed to foster elite engineers. They are rich in scientific research resources and have close school-enterprise cooperation. Their graduates are of high quality, can find good jobs with excellent remuneration. In recent years, new type education institutions has been established in France, and new type higher engineering education institutions represented by technical colleges and vocational colleges in universities have emerged.

On the basis of inheriting the featured elite education, France further put forth the concept "general-purpose engineer", and the corresponding curriculum provision fully reflects the comprehensiveness, diversity, flexibility, conformity and integration. Therefore, the disciplines and specialties are concentrated into groups of "engineering + management + economics". France adopts the five-year schooling system of "2 years preparatory work + 3 years of engineer training", and the average internship period is 18 months. France also pays high attention to the internationalized training of engineering talents, and through initiative actions and active promotion, it has established the European education pattern ("Sorbonne University Joint Statement"—*Bologna Declaration*—"European credit") required for the full implementation of EU integration.

6.3.5 China: CDIO Engineering Education Pattern

China has the largest higher engineering education system in the world. In 2015, 83% of the 2553 universities and colleges in China had engineering majors, with 1.18 million engineering undergraduates and 5.247 million students majoring in engineering in schools. In 2007, China launched the program to foster professionals short in key areas, giving priority to the training of engineering talents such as agriculture, forestry, water, geology, mining, petroleum, nuclear industry, software and microelectronics. In 2010, China launched the "excellent engineer program" and energetically advanced it during the 12th five-year plan period.

China applied to join the *Washington Accord* (WA) in 2013, and became a full member in June 2016. Joining WA is not only an important milestone in the internationalization process in the higher engineering education of China, but also started a new era in the world engineering education, especially for WA. In June 2016, the "UNESCO International Engineering Education Center", the only tier-II institution with engineering education as the theme of UNESCO, was inaugurated in Beijing, China.

In recent years, China has paid attention to the co-operative education of enterprises and schools, and vigorously promoted the CDIO engineering education pattern, namely, focusing on the whole life cycle of product conception (Conceive), design (Design), implementation (Implement) and operation (Operate), to allow students to study engineering in an active and practical way, with attention to the organic connection between different courses. Recently in particular, with joining the WA as an opportunity, China has deepened the teaching reform in engineering education, promoted internationalization of engineering education, established closer contact mechanism for engineering education with industrial enterprises, and also laid a solid foundation for establishing the registered engineer system.

6.3.6 Featured Engineering Education Patterns in Japan, Korea and Australia

Japan attaches importance to practice by students, and implements the training system dominated by "industrial laboratory" and "integration of industry-government-university". It pays attention to the close cooperation between enterprises and colleges and universities, and has developed the enterprise standard pattern of enterprise and university cooperation. Korea emphasizes that engineering education should approach to the fostering system of giving priority to technological innovation as required by demanders (the industrial circle), has set up the Accreditation Board for Engineering Education of Korea (ABEEK), practices the engineering education accreditation system, improves the enterprise-university-research-government cooperation system, established the Gwangju Institute of Science and Technology, Korean Advanced Institute of Science and Technology, the industry and university association, and the public technology transfer alliance, to promote the development and transformation of engineering technologies.

Australia is also unique in its higher engineering education pattern, enjoying a high reputation. Australia adopts rich teaching and learning patterns, the contents are updated quickly, and the students focus on independent study. In the learning process the "scenario" practical teaching is enhanced, and complete credit system is adopted with strict academic examination; it also focuses on strengthening team cooperation and innovation ability training through project curriculum.

6.4 Goals for Fostering Outstanding Engineering Talents

A general view of the status quo of the higher engineering education in the world reveals that problems are mainly concentrated in the following five aspects, which are particularly prominent in China today. The first is long-term discipline-orientation, contents-orientation and input-orientation, instead of result-orientation; the second is relatively enclosed training process, the connection, dialogue and interaction between the education and industrial circles are seriously lacking, and the practicality, applicability and innovativeness of engineering education need upgrading badly; the third is lacking in formalized, operable and internationally essentially equivalent assessment criteria for the knowledge and ability of students; the fourth is literalized engineering education, and it is imperative to strengthen the internship, practical drills, practice and operation links for students; the fifth is emphasizing on book-knowledge and ignoring abilities, emphasizing on mechanical and ignoring intelligence, emphasizing on hardware and ignoring software, emphasizing on unit (partial) and ignoring system (entirety), emphasizing on products and ignoring environment, emphasizing on technology and ignoring code of conduct, emphasizing on skills and ignoring ethics, and emphasizing on current and ignoring long-term, in short, there are prevailing phenomena of "emphasizing on hard aspects while ignoring soft aspects", and "emphasizing on things while ignoring people". Of these five problems, the first four are mainly with the state, authorities of education and education institutions, while the fifth is mainly with the students and teachers.

In face of the current profound changes in the higher engineering education in the world, how should we overcome the above problems, and especially, how to set the fostering goal for outstanding engineering talents in terms of student fostering goal? The following five aspects are key factors.

6.4.1 Systematic, Broad and Solid Scientific and Technical Basic Knowledge and Engineering Professional Skills

The basic knowledge of science and technology here includes the natural sciences such as mathematics, physics, chemistry and information science, as well as engineering and material sciences such as systems theory, cybernetics and engineering science. It needs to be emphasized that engineering education is based on engineering science and is supported by engineering technology, with the mission to foster excellent engineers who can turn scientific and technological innovation achievements into real productive forces. Unlike science education, engineering education pays more attention to engineering characteristics, and pays more attention to training and drilling on engineering professional skills and practical operation abilities, overcome the tendency of "literalized" engineering education.

Engineers must first master the knowledge and skills of engineering drawing, computation, experiment, testing, computer application and process operation, and can apply such knowledge and skills in solving practical problems. This is the primary feature of engineering and technical personnel different from other professionals. In addition, more comprehensive and in-depth high-level capability is needed. In fact, the CDIO training program divides the ability of engineering graduates into four levels of basic engineering knowledge, personal ability, interpersonal team ability and engineering system ability, the program requires that comprehensive ways of training be adopted, so that students meet the pre-set goals in these four aspects.

6.4.2 The Ability of Thinking to Become an Efficient Engineering Leader, Especially the Ability of Creative and Critical Thinking

Benjamin Bloom, a famous contemporary psychologist and educationalist, divided the education goals in the cognition field into six levels: knowledge, understanding, application, analysis, synthesis and evaluation. Usually, the first three categories and the latter three are respectively referred to as "low-level thinking ability" and "high-level thinking ability". To be a first-rate engineer, one must have high-level thinking ability, including: the critical thinking of analyzing, comparing, evaluating and predicting things; the practical thinking of answering questions, making decisions and proposing solutions; the creative thinking to imagine and create new ideas, solutions, or products; and the systematic thinking for overall planning and grasping of the urgency and interactivity, determining the priority and focus within the system, and trading off judgment and balance in making decisions.

The Massachusetts Institute of Technology (MIT), in the implementation of "Gordon ELP", put forward "six abilities of highly efficient engineering leaders", namely the leader character, comprehension-acceptance, meaning construction, goal vision, realization vision, technical knowledge and critical thinking. The MIT, through programs such as freshman curriculum, undergraduate practice opportunities scheme, engineering practice program, ELP short curriculum, ELP senior curriculum, engineering leader development program, Internship Plus program as well as engineering leader laboratory projects, trains the students on advanced thinking ability, to foster talents of engineering leader type.

6.4.3 Proper Mastery of Digitalized Technology and Intelligent Tools

From the traditional manufacturing characterized by mechanization and electrification, to the advanced manufacturing characterized by digitalization and networking, and further to the intelligent manufacturing characterized by collaboration and intelligence (powerful artificial intelligence and super artificial intelligence), the typical features of "reindustrialization" today are networking, digitalization and intelligentization, the basic routes are "Internet + industry", "Internet + engineering", and digitalized and intelligent engineering, the hot spots of competition are subversive and systematic innovation of key technologies and processes such as additive manufacturing, intelligent manufacturing, digital manufacturing, laser manufacturing, micro-nano manufacturing, and complicated manufacturing.

For example, in the field of construction engineering, the building information model (BIM) technology applies the information technology to realize the information management, sharing and interaction in the whole project life cycle of planning, operation and maintenance, leading to subversive transformation in the field of project construction. The first-rate engineering talents in the information age should master the digitalized technology and intelligent tools in interdisciplinary areas, and apply them actually in the engineering field to effectively solve engineering problems.

6.4.4 Healthy Personality, Professional Quality, Engineering Ethics and Humanistic and Artistic Accomplishment

First-rate engineers should pay attention to value shaping and personality cultivation, clearly know the engineer's roles and responsibilities, understand impact of engineering circle to the society and social formalization on the engineering circle, and have a clear view of history, cultural environment and global development perspective.

First-rate engineers should endeavor to acquire eight characters: engineering ethics, industrial integrity and professional quality; the sense of responsibility and compassion of caring for the environment and lives; a lifelong desire of learning and upward mobility to keep pace with the world engineering community; a self-conscious attitude towards life; a healthy personality with both persistence and flexibility; the innovation and entrepreneurship spirit to pursue for excellence and blaze with new trials and adventure; the craftsmanship of career veneration, earnest in work, responsible for products, constantly honing skills, persevering in the industry and studying in the field; the values of objective rationality and distinguishing between right and wrong, able to correctly understand the role of science and technology.

First-rate engineers should also strengthen humanistic and artistic accomplishment, and avoid becoming "robot" or "one-dimensional person". Correspondingly, universities should well combine the engineering education, professional education with general education, liberal education and innovation and entrepreneurship education. General education is the foundation of all education, and there is no exception in the fostering of engineering talents.

All foreign engineering colleges and universities pay attention to general education, in the University of Wisconsin-Madison, the proportion is quite high for general education in both courses setup and credits, even higher than those of professional class, especially in the humanities and social science education. The Worcester Polytechnic Institute specially set up the humanity curriculum system covering three categories of art, literature, and literal sciences, requiring that engineering students must take four courses of two knowledge groups and one course of another knowledge group.

6.4.5 The Comprehensive Quality of "Engineering with a Big E"

Fostering the students on the "engineering with a big E" was the dominating concept in the international engineering education today. The 12 quality requirements on graduates in the *Washington Accord* not only cover engineering knowledge and engineering ability, but also emphasize the general ability and virtue ethics, mainly including communication, team cooperation ability, and contents of social responsibility and engineering ethics.

Since the 1980s, the fostering of composite innovative engineering talents through interdisciplinary and dialogue forms has become an important part in formulating the engineering education strategy in developed countries. For example, the Ecole Polytechnique requires that students must choose courses of science foundation, engineering technology, economy, humanities, foreign language and sports, to cultivate their interdisciplinary comprehensive quality.

Norman Augustine, academician and chairman of the National Academy of Engineering, pointed out in the article *Engineering Education*: "industrial recruiting staff and administrative staff are always complaining that, graduated engineers are lacking in skills of written and oral communication, lacking in the ability to work as a team, lacking in the understanding of the business work to be carried out, and even lacking in the basic understanding of the political processes related to the existence or death of the project."

The good ability of communication, coordination and cooperation, being adept in oral or written (including writing communication, electronics and multimedia communication, graphic communication, etc.) publicity, explanation and persuasion, open international vision, the ability of cross-cultural exchange and communication, and successfully working and playing a leading role in a team, should all be covered in the fostering of first-rate engineers.

Chapter 7
Future of High-Speed Railway and University Education

7.1 Transition of HSR Industry and Technological Leap-Forward in "Post-HSR Era"

7.1.1 The Change of China's High-Speed Rail Industry in the Post-ministry of Railways Era

The "big transportation" system is a transportation management pattern widely used around the world, its main features are integration and general planning for railway, highway, waterway, aviation and pipeline, to form a five-in-one comprehensive, 3D and intelligent transportation system. The general background of China's revocation of the Ministry of Railways is the reform for separation of government from enterprises and "super-ministry system", with the aim of forming an efficient "big transportation" pattern.

According to the *Plan for Institutional Reform and Change of Functions of the State Council* deliberated by the National People's Congress on March 10, 2013, administrative functions would be separated from enterprise management for railway, and the management functions of the government, the society and enterprises will be defined and gradually separated through reform. According to this document, the administrative function of the former Ministry of Railways to work out railway development plans and policies would be transferred to the Ministry of Transport; the State Railways Administration would be organized under the Ministry of Transport, to undertake other administrative functions of the Ministry of Railways; and China Railways Corporation would be established, to undertake the enterprise functions of the Ministry of Railways.

F. Xu, *The Belt and Road*, https://doi.org/10.1007/978-981-13-1105-5_7

The revocation of the Ministry of Railways put an end of the powerful position of the railway sector, and HSR also entered the "post-Ministry of Railways era".[1] At the same time, the establishment of China Railway declared a new phase of reform for railway. As the main operation body of HSR was changed from the Ministry of Railways to the China Railway, it would face the following four major challenges:

The first is how to handle the relationship between sociality and economy, and between public welfare and profitability. In the era of the Ministry of Railways, railway provided quasi-public products, focusing on the sociality and public welfare, and the ticket prices and freight rate remained low in its operation, resulting to a certain extent in the low efficiency and heavy burden of railway transportation. After the establishment of China Railways Corporation (CRC), it is required to establish a corporate governance structure, participate in market competition, focus on changing the long-existing extensive management problems of railway such as emphasizing on input but ignoring output, low efficiency in assets application, poor cost control, and serious waste in accordance with the requirements of modern enterprise system, to effectively enhance its profitability.

The second is how to break the monopoly, centralization and enclosure, and fully stimulate the motive force of CRC itself and the vitality of local railway bureaus. On one hand, the former Ministry of Railways was a relatively enclosed system of its own, and the newly formed CRC, as a main market player, must fully arouse its internal impetus to face the market competition. On the other hand, the local railway bureaus under the CRC are also eager to obtain the power of dispatching commanding, transportation capacity distribution and autonomous pricing.

However, perhaps due to the inertia and tradition of the highly centralized and unified commanding for the whole country adopted all the time by the former Ministry of Railways, the restructured CRC did not (fully) delegate power although the administration organization was simplified, which has seriously restricted the vigor of local railway bureaus. The state must be determined to further carry out deep level reform of the railway system, to break the centralization of power. In the meantime, the local railway bureaus all over the country should be reorganized, to turn them into regional railway companies as corporate entities, and enable them to compete with each other in an orderly manner.

The third is reform in investment and financing. The CRC is a solely state-owned enterprise, the Ministry of Finance exercises the duty of investor on behalf of the State Council, and the Ministry of Transport and the State Railways Administration supervise it in the industry. However, the development of the CRC, especially the development of HSR requires widening the investment and financing channels. It is informed that during the 13th Five-Year Plan period, it has been planned to invest at least RMB 2.8 trillion yuan in railways, at an annual average of 500–600 billion yuan. With such a massive investment program, widening the investment and financing channels is all the more of primary importance in the reform of railway market.

---

[1] Long Quan: *How to Deploy Macro Transportation in Post-Ministry of Railways Era, Transport Business China* No. 4, 2013, pp. 64–67.

The inclusion of diversified capitals will surely demand reform of the existing institutional mechanism of the railway. The CRC should actively respond to the market changes, and make good use of diversified, socialized and marketized investment and financing forms such as public-private cooperation PPP and TOD, as well as advanced financing means of financing leasing, ABS (asset securitization), consortium loans and industrial investment funds, and encourage and expand civil capital, social capital and international capital to invest in HSR construction.

The fourth is technological innovation and intellectual property right. Innovation is an eternal theme. In fact, many very good innovations have been done by Siemens of Germany in active safety and modular car bodies, quality management system and reliability, in low resistance, lightweight, and aspects of disaster prevention and reduction by Kawasaki Heavy Industries of Japan, in ecological design by Alstom of France, and in energy-efficiency-economy-ecological (ECO4) by Bombardier of Canada, especially in technology innovation. What will the CHSR do?

The competition among countries in the world today is more than that in technologies, and it is more in technical standards, technical spectrum and technical paradigms, especially the competition in patents and intellectual property rights. In fact, the question of intellectual property right must be properly solved for the CHSR to go global. Even domestically, the question of intellectual property rights on HSR has become more complicated as compared with that before the restructuring of the Ministry of Railways, because now CRC is a corporate itself, and all aspects of enterprises, schools, research institutes and users has substantially enhanced their awareness of self-protection for intellectual property rights.

7.1.2 Technical Leap-Forward in the "Post-HSR" Era

The selection of technical paradigm has always remained a major strategic issue related to the development of HSR. Today, the wheel-track technology is adopted by HSR all over the world, as it is mature and with superior technical and economic performance to price ratio as compared with maglev. However, as pointed by academician Shen Zhiyun, the maglev High-Speed train will have obvious advantages when the ground travel speed is to be increased to over 400 km/h.[2]

The study of magnetic levitation technology was originated in Germany. In 1922, German engineer Herman Kemper proposed the principle of electromagnetic levitation, and applied for the patent of maglev train in 1934. There are two kinds of magnetic levitation: "normal conducting" and "superconducting". Further, normal conducting can be divided into "medium and low speed" and "High-Speed"; and superconducting into "low temperature" superconducting and "high temperature" superconducting. In working principle, superconducting levitation has bigger air gap than the normal conducting, and also higher travel speed, but the construction cost

[2]Shen Zhiyun: *Exploration on Technical Plan for Beijing-Shanghai High-Speed Railway, Journal of Traffic and Transportation Engineering* No. 2, 2001, pp. 10–13.

is also higher than that of normal conducting, therefore both have their advantages and disadvantages.

Compared with the wheel-track, maglev train using non-contact suspension technology has overcome the contact wear between vehicle and track, without the need to use high load-bearing rotating parts, and the propulsion system is built on the ground, therefore it has the advantages of low energy consumption, High-Speed, low noise, high safety performance, safe and comfortable riding, environmental protection and energy saving, wide adaptability, and small amount of maintenance. The biggest disadvantage of maglev is the high cost, even though, the progress of experimental testing for maglev train is still remarkable. The normal temperature short-stator maglev train of 80 km/h has been successfully developed by Southwest Jiaotong University. By adopting the German technology, the 430 km/h normal conducting long stator maglev train has been operating in Shanghai from Longyang Road Station to Pudong Airport. On April 21, 2015, the low-temperature superconducting maglev train reached a top speed of 603 km/h in the experimental line in Yamanashi Prefecture, Japan (Ozuki-Tsudome). In the research of high-temperature superconducting maglev train, Chinese research team has achieved the world's leading results.

The exploration and practice in vacuum pipeline transportation has also begun. The Tesla CEO Maske has set out to build "super High-Speed rail", this kind of super High-Speed rail is a futuristic rapid transit system, and can transport passengers in a low pressure pipeline at a speed of 760 mph (1220 km/h), close to the supersonic speed of 1224 km/h. The speed of this High-Speed train is faster than most civil passenger aircrafts.

According to a report on the *Daily Mail*, the Los Angeles based Hyperloop One Company, released on March 7, 2017 the photos of the test site for the super High-Speed rail (Pneumatic Tubes) called "DevLoop", the test site is located in the Nevada desert, and the first test is expected in the near future. DevLoop is essentially a long vacuum tube, and it will use pea pods (transport cabins) to transport people or goods, and its speed will be over 1000 km/h after completion. So far, Hyperloop One has raised $160 million to make DevLoop a reality. The system will be extended to the UAE and neighboring Gulf countries in the future, to greatly reduce the travel time between the two places. For example, it takes less than 50 min to travel between Dubai and Riyadh, as compared with two hours by airplane today.

The super-high speed rail is designed to overcome air resistance by using a vacuum or low pressure pipe. In fact, the highest economic speed of High-Speed transport running on the earth surface with dense atmosphere should not exceed 400 km/h, otherwise, 80% of the energy is consumed to overcome the air resistance, which is obviously not economic.[3] Some may question whether it is necessary for ground vehicles to reach a speed up to 400 km/h, as higher speeds have already been achieved long ago by civilian aircrafts. However, there are two insurmountable problems for aircrafts that rely on oil energy:

[3]Shen Zhiyun: *Dynamic Effect of High Speed Maglev Train on Tracks and its Comparison with Wheel-Track High-Speed Railway, Journal of Traffic and Transportation Engineering* No. 1, 2001, pp. 1–6.

One is that oil is a non-renewable source of energy, to depend entirely on oil, the mankind will soon face the question of energy depletion. In fact, competing for oil is the main field of power rivalry in the 20th century. In a sense, the chaos in the Middle East today has heralded the arrival of the post-oil era. Second, oil is a high carbon emission resource. The world's major developed countries have experienced the environmental pollution caused by excessive carbon emissions. Global warming is not the result of accumulation in a single day. Therefore, the development of more economical and environmentally friendly means of transportation is an inherent requirement from the sustainable development of human society.

Vacuum pipe transportation is expected to be a special option for future transportation. As the modern transportation of the 21st century, the development of vacuum pipe High-Speed and even ultra-High-Speed transportation should be promoted to the strategic height of the development of all mankind, and we must not be anxious for quick success and instant benefits.[4] Whether the wheel-track technology of CHSR can achieve further breakthrough, whether the maglev technology will be more adopted in the "post-HSR" era, how the vacuum pipe transport will be developed, and whether subversive technology paradigm transformation will take place in the near future, all these are worth our observation and expectation.

In the next generation of High-Speed rail equipment technology, the High-Speed train will have intelligent and adaptive bogie technologies, including the gauge-change technology required for HSR to go global, the light-weight car body technology with new materials, the high power electronic transformer technology, the permanent magnet motor and its traction control technology, full-electric braking technology, cophase power supply technology, energy saving transformer technology, equipment state monitoring technology, the application of new materials and new technologies, large capacity wireless communications and more accurate mobile blocking technology, intelligent transport organization and dynamic dispatching technology, and all these will be the core technologies for the development of the next generation of HSR. Furthermore, heavy loading technology with higher transport capacity, the renovation and upgrading technologies for existing railways will also be the emphasis of development of rail transport technologies.[5]

[4]Shen Zhiyun: *Exploration on Technical Plan for Beijing-Shanghai High-Speed Railway, Journal of Traffic and Transportation Engineering* No. 2, 2001, pp. 10–13.

[5]Fei Xu: *Speed up Making Breakthrough in Cutting Edge Technologies and Strategic Subversive Technologies in Rail Transit, Journal of the China Railway Society* No. 10, 2016, p. 1.

7.2 Mission and Journey of Research-Oriented Universities in the HSR Era

Research-oriented universities, as the cradle for training top creative talents, have natural education advantages such as rich education resources, complete rank of disciplines, excellent teams of teachers, the most talented young students, and free academic atmosphere.

Western developed countries relied on the first-rate universities to make the countries rich and strong and lead the world trend of development, this showed that first-rate universities always shoulder major missions and play an important role in rise of the country, national rejuvenation and social progress, industrial revolution, and cultural prosperity. The rise of universities of the UK in the 18th century, the surpassing by universities of Germany in the 19th century, and the leading by universities of the United States in the 20th century was all guided by fostering top-notch innovative personnel.

7.2.1 New Mission in the New Era: The Importance and Urgency of Fostering Innovative Talents

In the development of science and technology in modern times, the world has experienced four transfers of technology centers, and Italy, Britain, France, Germany and the United States became the world center of science and technology in succession, with the science and technology thriving period of each country about 80 years by average. There should be symbolic technologies to become a scientific and technological center and powerhouse, and where can such technologies come from? They are created by top-notch innovative talents trained in education. It will be difficult to produce symbolic technologies, become an innovative country with international competing power, and even more difficult to form powerful capability for sustainable development if top-notch innovative talents, leading and core personages are lacking.

Scholar Jiang Guojun pointed out in *On the Relationship between Education Center Transfer and Scientific and Technological Center Transfer* that, the prerequisite to become a world scientific and technological center is to become an education center. To implement the innovation-driven development strategy and build up a world scientific and technological center, and realize the leap-forward from following others to taking the lead in science and technology, it is imperative for China to foster a large batch of top-notch innovative talents. In October 2015, China formally issued the *General Program for Pushing forward the Construction of World Top-level Universities and Disciplines with Overall Planning*, specifically emphasizing to "speed up the reform of talent fostering pattern, promote science and education collaborated fostering, and complete the fostering mechanism of top-notch innovative talents to support high-level scientific research". The research-oriented universities in China must stand on the tide of the times, dare to shoulder missions, establish the scientific talent fostering value concept, rationally set the talent fostering goal for research-oriented universities, implement systematic reform of talent fostering pattern, and establish the system for top-notch innovative talents adaptive to a powerful scientific and technological nation.

Today, international competition is becoming daily intensified, the competition among countries are the competition of talents in the final analysis, and the competition of national strength and to take the first opportunities of development is focusing on talents, especially top-notch innovative talents. Speeding up the fostering of top-notch innovative talents has become a common strategic choice of world important powers to win initiative in the severe international competition. For the competition in the new century, many developed countries launched top-notch talent fostering programs and new deals for talents. In 1999, Korea began to implement the "BK21 Project"; in 2002, Japan started implementation of the "21st Century COE Program"; in 2005, Germany launched the "Excellent Program"; in 2006, Russia implemented the "Federal Innovative University Program"; in recent years, the U.S. government has again approved the "American Competitiveness Program" designed to attract the world's top-rate students and maintain the attractiveness of U.S. higher education. Australia has begun to quietly cut course spending and introduce scholarship incentives to attract outstanding overseas students. The European Union has issued the "European 2020 Program" to increase investment in the research and development area to 3% of GDP, to attract and introduce more high-level excellent talents.

The 21st century will see the rapid development of science and technology, new technology, new energy and new materials keep on emerging, cutting-edge sciences and technologies such as a new generation of information and communication, aviation and aerospace, biomedicine, intelligent manufacturing, quantum communication change with each passing day, the development of artificial intelligence and brain inspired robot is especially impressive, and the mankind has entered the modern age to play together with artificial intelligence. In the man-machine Weiqi duel attracting the world attention in March 2016, the AlphaGo beat the world Weiqi champion Lee Se-dol 4:1, another milestone in artificial intelligence after the "IBM Deep Blue" beat the world chess champion in May 1997 with two wins three even and one loss in six games.

On January 4, 2017, as the Chinese chess player Gu Li admitted defeat in the 30 s for one throw blitz confrontation, the Master AlphaGo (latest version) ended with winning all 60 games, more than half the Weiqi world champions were defeated in their challenges, and dozens of top go players of China, Japan and Korea including Nie Weiping, Ke Jie, Park Jeong-hwan, Yuta Iyama were all defeated. If we do not quickly foster innovative and creative talents, it is not scaremongering that the artificial intelligence of machines will surpass the intelligence of people, and this is also another dimension to understand the importance and urgency to foster top-notch innovative talents.

7.2.2 *New Models and New Journey: The Challenges and Arduousness in Fostering Top-Notch Innovative Talents*

Since the mid-20th century, a series of creative experience has merged in foreign countries, such as the general education curriculum in Harvard University, the Melbourne model in the University of Melbourne, the research-oriented education or research-based education and entrepreneurship university construction in education-oriented universities in Europe.

The world higher education is embarking on a new surge of reform, and the concept and model of higher education is undergoing fundamental changes in its various fields. At present, the important demonstration of the new pattern of higher education is the construction of system to foster top-notch talents that integrates general education, professional education and innovation education. Columbia University, Harvard University, Yale University and other top universities in the world have re-examined the undergraduate education under the new pattern, and reinvented the undergraduate education with multiple measures. Duke University, ranking at the 8th place in the "U.S. News 2015 Ranking of American Universities", pushed forward the reform of university undergraduate education by rebuilding it from the very beginning with large amount of investment, and launched the general education based on "Duke immersion learning", and it did overall renovation from the specific courses to the whole curriculum system, teachers configuration, extracurricular practice, and even the architectural structure of students' dormitory.

Globalization and digitalization is another important feature of the new pattern of higher education. The tide of globalization has already promoted a new round of talent training pattern reform. A new round of global scientific and technological revolution and industrial transition has become the basic driving force of education reform in many countries of the world. As Merritt Jeno, dean of the School of International and Public Utilities of Columbia University, said, "how to further participate in the global education, and, in turn, influence the global thinking, has become the foundation of a good university".

At the same time, informatization is sweeping across the whole world, the human society is fermenting the fourth industrial revolution marked by digitalized production, digitalized means such as cloud computing, big data, intelligentization and mobile Internet will be widely used in different fields, and the education sector will also undergo subversive transformation and structural changes due to digitalization. With the development of teaching ideas such as "cooperative learning" and the application of emerging education and teaching methods represented by MOOC, the traditional education patterns have been strongly impacted, new education patterns such as MOOC, flipped classroom, blended teaching, smart campus, and digitalized university are highly praised by the people due to their promotion of diversity, openness and personalized learning to the maximum extent.

Faced with the multiple challenges of globalization and digitization, Chinese universities, especially research-oriented universities, have actively explored and practiced on fostering top-notch innovative talents in the "trinity" pattern. College students innovative experiment plan, talent training pattern innovative experimental area, graduate student education innovation plan and basic discipline and applied discipline top students test plan keep on emerging, and the "Yuanpei Academy" of Beijing University, the "Tsinghua School" of Tsinghua University, "Zhiyuan College" of Shanghai Jiaotong University, the Four-Year Boarding Academy" of Fudan University, the "Zhu Kezhen College" of Zhejiang University, the "Kuang Yaming College" of Nanjing University, and the "Chongshi Academy of Xi'an Jiaotong University and so on, are showing their increasingly exuberant vitality. However, such exploration and practices are still preliminary, phase-specific and partial, and are still far from the goal of forming the system to foster top-notch innovative talents, therefore universities in China, especially those research-oriented still have a long way to go with heavy tasks.

Over the years, Southwest Jiaotong University, Beijing Jiaotong University and other research-oriented universities with distinctive feature of railway transportation and powerful strength, have spared no efforts in exploring how to foster top-notch innovative talents. The great development of CHSR also provided these schools with rare and precious opportunities. China High-Speed Railway has become important facility of a great power and global business card. It is the painstaking efforts of the countless outstanding talents that have been trained by the railway higher education behind this world attractive name card.

In a sense, the development road of China's rail transport industry from scratch, from weak to strong, and from following others to taking the lead, is a process of fostering the top-notch innovative talents for rail transport in China, and also a road of exploration to run universities with distinctive features of rail transport. At present, universities and colleges with features of rail transport should continually consolidate and expand their features and advantages in transportation, take initiative to shoulder the historical mission of fostering high-end talents and leaders who can lead the future development for the coming HSR era and "post-HSR" era, work hard to create the updated version to foster top-notch innovative talents in new opportunities, and speed up the construction of first-rate research-oriented universities with distinctive features of transportation.

7.3 Higher Education for Railway in China and Roles of Universities

7.3.1 Development Road of the Higher Education for Railway in China

The higher education for railway in China was started in the Qing Dynasty in the late 19th century and is continuing today. Its development can be roughly divided into four stages since the 1950s.

In the first stage, from 1952 to 1976, some universities and colleges were designated as industry schools in the initial period after the establishment of New China and the subsequent school readjustment, to foster and reserve professional talents for the railway cause of New China.

In the second stage, from 1976 to 2000, the former Ministry of Railways of the People's Republic of China was in charge of the higher education for railway, and the development pattern was formed of three echelons of general special schools, ordinary undergraduate schools and key schools for both academic and non-academic education, with a total of about 40, to constitute the university group supporting the construction and development of railway and a complete system to foster and train professional railway talents together with a small number of universities and colleges under other ministries of the government and associated with railway.

In the third stage, from 2000 to 2013, in the surge of school readjustment and merging in China, the key schools of railway were allocated to the Ministry of Education. After that, most of the universities and colleges well retained the features of railway industry, and continually provide powerful disciplinary and talent support to the development of railway undertaking.

In the fourth stage, starting from 2013 up to date, with the rapid development of High-Speed railway in China and in the world, the higher education for railway of China has been informatized, digitalized and deeply internationalized, they carry out all-round international exchange and cooperation to upgrade the education and teaching quality comprehensively, and conduct practical and overall combination in technology, management, design, construction and testing of railway system on aspects of discipline, specialties and scientific research, well matching the talent and scientific and technological demand on the railway market, especially HSR market of China.

The China High-Speed Railway (CHSR), with the unremitting hard work by people of generations, and the powerful support by the railway universities and colleges, has achieved the historical leap-forward from scratch and from weak to strong, and experienced the process from manufacturing to creation, from exploration to breakthrough and from catch-up to taking the lead in the course of independent development of test trains, import to obtain mature experience on trains, and then self-reliance research and development of totally new trains on the basis of new experience. Today, HSR has not only become a national name card of China, but more a symbolic area of achieving major breakthroughs in the construction of an innovative country.

In general, the higher education for railway of China has provided technology and talent support to China railway; in turn, the substantial development of railway demand, especially the speed increase of China railway and the great development of China railway in recent years has also created unprecedented development space for the railway schools. The higher education for railway and railway construction in China have promoted, pushed ahead and benefited from each other.

7.3.2 Belt and Road and Undertaking by Universities and Colleges

As described above, the Belt and Road run across the Asia, Europe and Africa continents, with one end being the active East Asian economic circle, the other the developed European economic circle, and a hinterland with huge potential for economic development between them. These economic features of the Belt and Road are known to all. However, what can be ignored is the sources of oriental and western civilizations at both ends of the Belt and Road and the cross-culture exchanges along the way, and this is just the central role that universities and colleges should play in the Belt and Road. It is the responsibility, undertaking and mission of universities and colleges to push forward the exchange, communication and integration of countries along the Belt and Road. Universities and colleges can play an irreplaceable role in jointly upgrading the internationalization level of education and the capability to serve the Belt and Road, to make new and unique contribution to the "community with a shared future for mankind".

In fact, education has the basic, pilot and supporting role in the great historical cause of joint construction of Belt and Road. In July 2016, the Ministry of Education of China issued the *Education Action to Push Forward Joint Construction of Belt and Road*, emphasizing that education is fundamental to national prosperity and people's happiness, putting forth the principle of "fostering people as fundamental, humanity taking the lead, government guidance, people as main player, joint consultation and construction, opening and cooperation, harmonious and inclusive, and win-win mutual benefit", calling for the joint construction of Belt and Road education community (Fig. 7.1).

Strengthening the exchange and cooperation of countries along the lines in education, especially cooperation in higher education, should certainly become an important component in the joint construction of Belt and Road education community. At present, universities and colleges in China have established humanistic exchange mechanism in different forms with universities and colleges in countries along the Belt and Road, such as China-Russia university president summit, China-Egypt university president forum, China-Israel university president forum, China-ASEAN experience exchange week, and China-Africa higher school "20+20" cooperation program. Some universities and colleges in China also established overseas cooperative education in different forms with universities and colleges in countries along the Belt and Road, and "going global" and "introduction" also show good situation.

As actions to go global, Xiemen University set up the Malaysia Branch, Beijing Jiaotong University set up China-Russia Transportation College, Suzhou University established Laos Suzhou University, and Southwest Jiaotong University is speeding up the joint construction of India Railway University and preparing for the Ethiopia Railway Research Institute, and so on. In "introduction", just take Shenzhen as an example, Moscow University and Beijing Institute of Technology have cooperated in running a school in Shenzhen. We can say that higher education cooperation is playing an increasingly important role in promoting dialogue and exchanges of Chinese civilization with other exotic civilizations, and promoting the mutual understanding and people-to-people bond in countries along the Belt and Road.

The education in countries along the Belt and Road has distinctive features, rich resources, high complementarity and huge space for cooperation. Universities and colleges of China should, in the concept of "inclusive development", actively link with the universities and colleges in countries along the way, learn from each other the advanced experience in education, share high quality education resources and promote the accelerated development of higher education in all aspects. China should work hard in the connectivity cooperation in education, further strengthen the cooperation and exchange with universities and colleges in countries along the Belt and Road, make more efforts in "going global" and "introduction", and carry out pragmatic cooperation at multiple levels and in multiple fields. We can, backed up by the disciplinary advantages of our own, unite with relevant universities and colleges along the ways to jointly construct international cooperation united laboratories combining production, learning, research and application, actively build national and regional research bases, and carry out researches in economy, politics, education and culture areas with universities and colleges in countries along the Belt and Road.

In the meantime, we should endeavor in talent fostering and training cooperation. Domestically, we will foster composite senior talents good at both engineering and management, proficient in both language and commerce, and in both legal affairs and

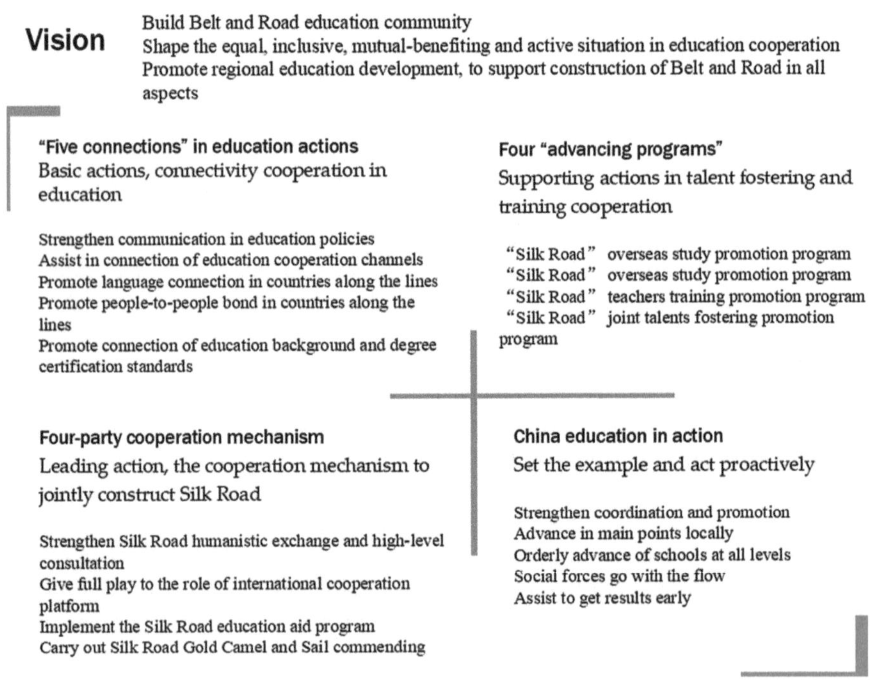

Fig. 7.1 Education action to push forward joint construction of belt and road

culture; for countries along the Belt and Road, we can arrange more students to study in China, and improve the training quality for these students. In cooperation with the going global of HSR, nuclear power and communication enterprises of China, we will explore cooperative education in various forms in countries along the Belt and Road, and arrange professional training at multiple levels, to jointly train students in professional areas in urgent need.

Transportation changes the life and education achieves the future. It is the duty of railway related universities and colleges to foster HSR talents with high accomplishment, high level and rich humanistic feeling. While focusing on professional and skill training, higher education should well shoulder the mission to promote the human civilization and culture exchange and communication and enhance the human welfare. It should be emphasized that, HSR should not be merely the patent of economically developed countries and regions, universities and colleges, research institutions and related enterprises in rich and advanced countries should help underdeveloped countries and regions to build High-Speed railways at appropriate time via open research and experiment platforms, establishing joint R&D centers and aids, to promote the inclusive development of the world.

7.4 Fully Support the HSR R&D and Construction: Exploration and Practices by Featured Research-Oriented Universities

Southwest Jiaotong University, founded in 1896, now has a glorious history of 121 years across 3 centuries. Southwest Jiaotong University was born and grew up for railway, and became powerful with High-Speed railway. It is the originating place of the modern higher education of China in civil works, transportation engineering and mining and metallurgical engineering, and the cradle for railway engineers and HSR talents of China.

The long time hard work and outstanding contributions of Southwest Jiaotong University has vividly annotated and sublimated the value pursuit and lofty ideal of developing transportation for cultural exchanges with the aim of great unity. In recent years, the school has speeded up the construction of a first-rate comprehensive research-oriented university with distinctive feature of transportation, to endeavor to upgrade and demonstrate the scientific and technological creativity, academic competing power and ideological influence of a century-old school.

With long efforts, Southwest Jiaotong University has built up the three systems of disciplines, talents and scientific research that support the modernization construction of rail transport in China and economic development in the served areas, shaped the comprehensive advantages with the longest school history, fullest rank of disciplines, most complete professional association and most concentrated core resources in the rail transport field, and has provided and will continue to provide powerful support to the research and development and construction of High-Speed railway of China.

7.4.1 Talent Fostering and Training

Talent training, scientific research, serving the society, cultural inheritance and innovation, and international exchange and cooperation are the five major missions of universities, and talent training is the most fundamental of all. Talents are the inexhaustible motive force and important support to the development of HSR, and universities and colleges should serve as talent fostering bases with their profound scientific research strength and rich intelligence resources. As railway comes to the HSR and post-HSR era, talents fostering and training has become more prominent as the foundation and pilot work.

For over 100 years, Southwest Jiaotong University fostered many scientific and engineering masters shining in history such as Mao Yisheng, Lin Tongyan, Huang Wanli and Zhu Kezhen, and over 300,000 talents represented by 59 domestic and overseas academicians, and almost all academicians in the rail transport field of China graduated from Southwest Jiaotong University.

The school carries on the tradition of serving railway transportation, strengthens collaboration of enterprise, school and research, and spares no efforts in fostering high level talents needed in rail transport. To meet the general trend of reform in world higher education and the pressing need from the rapid development of HSR, the school developed a number of international courses including "High-Speed Railway Engineering", published relevant teaching materials and works all in English, and planned and developed more than ten HSR MOOC courses, which are all online.

The school has carried strategic cooperation with UIC, has sent doctorate graduates to the UIC headquarters for joint fostering and practical training in four consecutive years, and sponsored the "International Rail Transport Academic Forum" together with the UIC a number of times. On the basis of successfully conducting the high-end training programs for railway projects in Ethiopia, Iraq and Uganda, it is now engaged in the training of technical and management talents needed by countries along the China-Europe railway, and it has set up the "SWJTU International Training Center on China Railway Technology", to provide package talents support for the HSR to go global.

7.4.2 Scientific and Technological Innovation and Taking the Lead

For a long time, SWJTU has adhered to the specialty of transportation in the university education. From the birth of the first manned high temperature superconducting maglev experimental vehicle in the world to the completion of the first vacuum pipe high temperature superconducting maglev traffic test line in the world; from the research and development of the virtual track tram to the proposal of transportation control system of hybrid right of way tram based on the technology of Internet of vehicles, all these have demonstrated the strong innovation awareness and firm willpower to lead the development of rail transport of the SWJTU teachers and students.

In the epoch-making projects of Yunnan-Myanmar Highway, Baoji-Chengdu Railway, Chengdu-Kunming Railways, Qinghai-Tibet Railway, and Beijing-Shanghai High-Speed Railway, and mass transit projects of medium and low speed maglev and high temperature superconducting maglev trains, Southwest Jiaotong University has created more than one hundred Number One's of China and the world, steadily ranking at the first place in the traffic and transportation discipline. A large number of major scientific and technological innovations with independent proprietary intellectual property rights have become the main brands for CHSR to go global.

Professor Cao Jianyou (academician of the Chinese Academy of Sciences) proposed a single-phase AC 25 kV power supply system for electrified railway, setting the direction for the development of the railway electrification in China. Professor Qian Qingquan (academician of Chinese Academy of Engineering) opened the research direction of railway microcomputer monitoring and integrated automation. The cophase power supply technology proposed by Professor Li Qing-Zhan and the wound core energy saving transformer developed as led by Professor Gao Shibin, have become the new technology direction of the future railway transportation traction power supply. Professor Shen Zhiyun (academician of Chinese Academy of Sciences and Chinese Academy of Engineering) proposed functions of the low power wheel track, indicating the direction of the heavy loading and High-Speed bogie.

The High-Speed train coupling system dynamics theory developed on the basis of the "Shen Theory" of Professor Shen Zhiyun and the "Zhai-Sun Model" of Professor Zhai Wanming (academician of Chinese Academy of Sciences) and Professor Sun Xiang has become the direction of the vehicle system dynamics research. In the direction of basic research and innovative development of High-Speed trains in China, Professor Zhang Weihua and his team have put forward several innovative points in the *Overall Technical Conditions of Motor-Train Units of* 300–350 km/h.

SWJTU has solid scientific research foundation and competence in the development of HSR. At present, it has 11 national laboratories, key laboratories and engineering research centers (laboratories), international scientific and technological cooperation bases, and many provincial and ministry level scientific research platforms closely related to transportation, and has also constructed a number of HSR soft scientific research platforms including "China Strategic Research Center for High-Speed Railway", "TOD Research Center", "PPP Research Institute" and "China HSR Intellectual Property Rights Research Center". The 11 national scientific research platforms are: Rail Transit National Laboratory (Preparatory), National Key Laboratory on Traction Power, Rail Transportation Electrification and Automation Engineering Technology Research Center, National Engineering Laboratory for Land Transport Geological Disaster Prevention Technology, National and Local United Engineering Laboratory for Intelligenization of Integrated Traffic and Transportation, National and Local United Engineering Laboratory for High-Speed Railway Operation Safety Space Information Technology, International United Research Center for Modern Transportation, Communication and Sensing Network, National and Local United Engineering Laboratory for System Reliability Automatic Validation, National and International Science and Technology Cooperation Base for Modern Rail Transit Vehicle Design and Safety Assessment Technology, National Engineering Laboratory for Urban Rail Transit System Safety Assurance Technology, and National Engineering Laboratory for Integrated Traffic Big Data Application Technology.

Science and technology innovation is the driving force of the future development of High-Speed rail, the reality basis for the "faster, smarter, safer, more economical, more environmentally friendly and more humane" High-Speed railway in the future, and also the fundamental guarantee to achieve technological breakthrough and leap-forward for High-Speed railway in the future. SWJTU adheres to integration of science and teaching, and has carried out deeply the basic research, core technologies and equipment innovation and integrated testing and verification for High-Speed railway by relying on the "Rail Transport Safety Collaboration Innovation Center" in the first batch of national "2011 Program" constructed with the school taking the lead and a number of outstanding scientific research platforms, to provide innovation results and technical support to the CHSR going global.

In 2015, the rail transport national laboratory under construction set the world record of 600 km/h in the vehicle rolling vibration bench test. At present, SWJTU is carrying out innovative researches on HSR, heavy loading and new type urban mass transit, High-Speed and ultra-high speed high temperature superconducting maglev transportation in four aspects of development strategy, basic science, support technology and cutting-edge technology for the national strategies of Belt and Road, Made in China 2025 initiative and HSR going global, striving to realize breakthroughs and leap-forward in key, common and cutting-edge technologies.

7.4.3 International Exchange and Cooperation

In the trend of the High-Speed development and globalization of world economy, in the future, High-Speed rail will be one of the most basic forms of road transportation between countries around the world, and one of the most main ways for connectivity between nations worldwide. It is quite necessary to conduct international exchanges and cooperation in the field of High-Speed rail, by international exchange and communication, learning from each other and complementing each other among peers including universities, research institutions and enterprises, so as to obtain better quality for High-Speed railway at lower cost, higher efficiency and in shorter period of time.

Education internationalization strategy is one of the three strategies set by Southwest Jiaotong University. Since 2010, SWJTU has established partnership with the International Union of Railway (UIC), the University of Illinois at Urbana-Champaign (UIUC), Taylor Francis Publishing Group and Elsevier Group. In particular, it has set up a High-Speed railway education, scientific research and development center jointly with the UIUC, it is the only university strategic partner of the UIC in

China, and has sponsored jointly three sessions of rail transit international academic forums.

The school has carried out short-term international railway training programs for Ethiopia, Uganda, Kenya, Tanzania, Iraq, Vietnam, Russia, Pakistan and Laos. In May 2015, the university, as recommended and designated by the Chinese Government, cooperated with the Indian Ministry of Railways to help build the Indian Railway University. In 2016, the school actively promoted the comprehensive strategic cooperation with Russian universities, especially railway universities in the field of rail transport.

Appendix

A.1 Adhere to the Road of Indigenous Innovation with Chinese Characteristics[1]

—Successful practices of HSR of China

At the National Conference on Science and Technology Innovation, the Conference of Academician of Chinese Academy of Sciences and Chinese Academy of Engineering and the Ninth National Congress of China Association for Science and Technology, General Secretary Xi Jinping called for the construction of an innovative country and a powerful nation of science and technology in the world, he stressed that "we must adhere to the road of independent innovation with Chinese characteristics, face the cutting edge of world science and technology, the main battlefield of national economy and the major demand of the state, speed up science and technology innovation in every field, and seize the first opportunities in global supervisory control system competition". At the conference, General Secretary Xi Jinping fully affirmed many engineering and technological achievements including High-Speed railway, and pointed out that these results, together with other achievements and breakthrough in basic sciences, have laid an important foundation for China to become a big nation with world influence. The development of CHSR has experienced a process of starting from scratch, growing from weak to strong, and from "following the others" to "running side by side" and then "taking the lead", realized the leap-forward surpassing from "technology import" to "made in China" and then to "created in China", and has established a complete system of survey, design, construction, equipment, operation and safety management standards for HSR, the brands and proprietary intellectual property rights for HSR equipment, and has become a benchmark of "Made-in-China" and "China speed". CHSR has successfully opened up a road of innovative development in HSR with Chinese characteristics, and is successful practice in taking the road of indigenous innovation with Chinese characteristics.

---

[1]This is the signed article published by the author of this book on *Seeking Truth* No. 4, 2017.

F. Xu, *The Belt and Road*, https://doi.org/10.1007/978-981-13-1105-5

A.1.1 The Innovative Development Course of CHSR

Since the first Beijing-Tianjin Intercity High-Speed Railway with a speed of 350 km/h was opened for traffic in 2008, China has completed the HSR with the highest grade in the world the "Beijing-Shanghai High-Speed Railway", the first HSR in high and cold area in the world "Harbin-Dalian High-Speed Railway" and the world longest HSR "Beijing-Guangzhou High-Speed Railway" (2298 km). By the end of 2016, the operation mileage of HSR in China exceeded 20,000 km, ranking the first in the world, and accounting for about 65% of the total mileage of HSR in the world. In 2015 China also reached agreements with Russia and Indonesia, to build in cooperation the Moscow-Kazan HSR and Jakarta-Bandung HSR, successfully realizing the breakthrough progress for the CHSR to march from domestic to the world market.

CHSR has obtained great achievements, won extensive praises in all walks of life, and has been fully confirmed by the leaders of the CPC Central Committee and State Council. After inspection of relevant enterprises, General Secretary Xi Jinping pointed out that, "HSR is a bright name card of equipment manufacturing of China". Premier Li Keqiang emphasized at different occasions that CHSR has the great advantages of advanced technologies, good safety and reliability, high performance to price ratio and rich operation experience. The innovation course of CHSR can be roughly divided into four periods: technology accumulation period (up to 2004), the period for technology import, digestion, absorption and re-innovation (2004–2008), the indigenous innovation period (2008–2012), and all-round indigenous innovation period (from 2012).

Technology accumulation period: a solid foundation was laid for the innovative development of HSR. Early at the beginning of the 1990s, the former State Commission of Science and Technology included the "complete set of HSR technologies" into the state key scientific and technological program of the 8th Five-Year Plan period, and organized relevant scientific research forces to work jointly on key technical issues in HSR, making the necessary initial technical reservation for the subsequent construction of Beijing-Shanghai High-Speed Railway. After that, the Chinese railway sector, with their own technical forces and indigenous innovation, developed more than 10 types of high speed trains represented by "China Star". Although there was a fairly big gap between the High-Speed trains developed in China and those in advanced foreign countries, China provided the technology and talent reservation for the technology import, digestion, absorption and re-innovation in the subsequent period through independent research and development.

The period of technology import, digestion, absorption and re-innovation: it injected new motive force for the innovative development of HSR. In April 2004, the State Council issued the *Minutes of Meeting on Relevant Issues in Research of Railway Locomotive and Vehicle Outfitting*, making clear the technical lines for the import, digestion, absorption and re-innovation of High-Speed power train units, and the former Ministry of Railways and the National Development and Reform

Commission decided to implement the action plan "with overall technology transfer as the prerequisite, the import of core and key technologies as the main point, the domestic enterprises as the main bodies, and localization as the ultimate goal". China imported the world most advanced High-Speed motor train units (MTU) manufacturing technology, successfully built the technological platform for High-Speed trains of two speed levels of 200 and 300 km, and successfully mastered the nine core technologies and ten main associated technologies of High-Speed MTU. This process injected new motive force for the innovative development of HSR in China, greatly pushed ahead the overall progress of HSR in basic theory and research of key technologies, and substantially upgraded the technical outfitting level of HSR.

Indigenous innovation period: powerful support was given to HSR to form the proprietary intellectual property rights. In 2008, the former Ministry of Railways and the Ministry of Science and Technology jointly signed the *United Action Plan on Indigenous Innovation for High-Speed Trains in China*, which stated to establish and complete the CHSR technology system with proprietary intellectual property rights and powerful international competing power with speed of 350 km/h and over. The two ministries started united actions, to realize scientific and technological innovation in the form of the government taking the lead and enterprises acting as the main body, with market as orientation and project cooperation as ribbon, and push forward the positive design and indigenous innovation for High-Speed trains in China. In 2010, the China-made CRH380 series High-Speed MTU, with high degree of proprietary intellectual property rights, was put into service, successfully realizing breakthroughs in key technologies of head-to-head, lightweight vehicle body, bogie, vibration and noise reduction and system integration, laying a solid foundation for realizing overall indigenous innovation of core technologies.

Period of overall indigenous innovation: this period pushed the innovation of HSR into "Chinese standard". In 2012, according to the innovation-driven development strategy of the state, China actively pushed ahead the development of MTU to Chinese standards with full proprietary intellectual property rights in the principle of adhering to proprietary intellectual property rights, safety and reliability, standardization, serialization, simplicity and unification, economic performance, energy saving and environmental protection. In June 2015, the first standard MTU of China was officially completed, marking that the innovation of HSR officially has stepped into the era of "Chinese standards". The technical standard system for Chinese MTU, while adopting many national standard, industrial standards and special technical standards of China, has also actively integrated with some international standards and advanced technical standards of foreign countries. Also, with the export of High-Speed MTU as breakthrough, the export of complete set of HSR technologies for construction engineering, traction and power supply, communication signal and operation management was started, to build the "CHSR" brand and realize the strategic goal of CHSR "going global".

A.1.2 Key Factors of Successful Innovation of CHSR

In a short period of a dozen years, while growing into the largest HSR market in the world, China has also become an important supplier of rail transport equipment in the world, having scored extraordinary achievements and development speed. Why has the indigenous innovation of CHSR become successful? There are the following key factors.

The first is proper top-level design and overall planning with the government taking the lead. The Chinese Government did top level design in good time and effective overall industrial planning on the basis of the conditions of the country and its railway, plotted the development blueprint for CHSR with "four longitudinal and four transversal lines" at the national strategic level, and set the technical line of "import, digestion, absorption and re-innovation" for the locomotive and vehicle equipment for CHSR. The competent department of railway integrated the domestic market with strong power, and started negotiations with foreign companies with unified principles. By designating strategic buyers and setting the import principles that "critical technologies must be transferred, the price must be the lowest, and Chinese brands must be used", not only avoiding the possible awkward situation that the domestic enterprises are conquered one by one, but also breaking the technical "moat" intentionally held by the powerful HSR countries from an advantageous position and at low cost, promoting the domestic companies to dock with the international most advanced technology platform at high level, and laying the solid foundation for indigenous innovation by CHSR at a high start point.

The second is giving full play to the role of enterprises as main body. Enterprises are the main bodies of market and industry, and more the main bodies of technological innovation. The HSR enterprises of China made full use of the market as the decisive force that allocates the innovation resources, to push the rapid development of HSR. The first is to guide the innovation direction of HSR enterprises as directed by the market demand. The CHSR had to carry out targeted R&D and innovation of HSR technologies with the complicated geological and climate conditions of China and the operation technology demand of long distance, high density and different speeds for operation on the same line or across lines. The second is to increase the indigenous innovation capability of HSR enterprises aiming at market competition. Orderly competition in the industry was carried out on the basis of market principles with the former CSR and CNR and three main technical platforms as the representative, to realize indigenous innovation in HSR. The third is to lever the innovative development of the whole industrial chain with industrial collaboration as the motive force. HSR, as a knowledge and technology intensive industry, its innovation is interdisciplinary, cross-specialty and cross-industry complex system engineering, so it is necessary to deploy the innovation chain for the industrial chain, to promote the innovative development and overall upgrading of the whole industrial chain of HSR.

The third is to fully rely on the scientific and technological and talent resources of integrated enterprise, school and research. It is the key to successful innovation of HSR to give play to the comparative advantages of universities and colleges, scientific and technological institutions and enterprises on the basic research, application research and development research, bring into full play the main forces of science and technology and talents, and build the open innovation platform combining enterprise, school and research with market demand as orientation and the scientific research projects as backup. The scientific and technological innovation pattern of enterprise, school and research integration has enabled to not only make efforts to break major key technologies for the present to support the rapid development of the HSR industry, but also deploy in advance cutting-edge technologies for the future, to lead the sustained and coordinated development of the HSR industry in the future, as powerful support to the innovation in CHSR.

The fourth is to energetically carry on the HSR spirit of dedication, pioneering and pursuit for excellence. Spirit is the source of force and the pilot of action. For such great achievements scored by CHSR in a short period of a dozen years, the role of mental factor cannot be neglected. The "locomotive" spirit and "Qinghai-Tibet Railway" spirit of the railway people, combined with the railway accomplishment of taking the state interest as the most important, bold to take responsibilities, diligent in learning and practice work and strive for innovation, has concentrated into the powerful mental strength of patriotic devotion, dare to pioneer, and pursuit for excellence of the railway people, and become the precious mental wealth to realize the innovative development of CHSR.

A.1.3 Beneficial Inspiration in the Innovative Development of CHSR

Conscientious summary of the successful experience of CHSR in innovative development can well inspire us today in deepening the reform of scientific and technological system, implementing the innovation-driven development strategy and building the innovative country, and is of important reference significance to other industrial sectors, especially the high-end equipment manufacturing industry with huge market size and complicated product systems.

Inspiration I: give full play to the advantages of the socialist system and concentrate forces to accomplish major tasks. General Secretary Xi Jinping pointed out: "Our greatest advantage is that the socialist system of China can concentrate forces to accomplish major tasks." The High-Speed rail technology of China features great complexity and high comprehensiveness, it needs the close cooperation of all sectors. The *Medium and Long-term Railway Network Planning* approved by the State Council in 2004 planned the strategic blueprint for the development of railway transport in China at the national level, and guided by this plan, the competent authorities of railway imported advanced MTU technology from other countries at low cost by institutional means of unified market, unified negotiations with foreign companies and unified product procurement, and formed the situation of national unified arrangement to deal with key problems with enterprises as the main body and the enterprise, school and research cross-sector collaborated innovation as the motive force, fully demonstrating the highly effective organization and mobilization capability, authoritative commanding and coordination power and strong action force with unified steps, and fully reflecting the advantage of socialist system that forces can be concentrated to accomplish great tasks.

Inspiration II: firmly take the road of indigenous innovation, and maintain the strategic willpower of taking ourselves as the main. The important reason that the CHSR could surpass others in a short period of time is the full utilization of global innovation resources, and practices and exploration aimed at realizing indigenization of core technologies. In the era of globalization, the utilization of global innovation resources becomes more convenient, however, core technologies, especially resources that can continually raise the competing power of an industry have become even scarce, difficult to obtain through transaction on open market and it must rely on indigenous innovation. The hard work history of CHSR has shown that, the railway people adhered to the awareness of indigenous innovation in all periods of development, and moulded the real innovation self-confidence with the strategic willpower of taking ourselves as the main. China adopted an important policy in importing technologies at that time: all parts must be made in China, even for some parts that must be brought abroad as they could not be made in China for a time being, the foreign company can be included in the list of suppliers only when it has formed joint venture with domestic enterprise. Such policies are vivid interpretation of the awareness and willpower of indigenous innovation.

Inspiration III: Aim at "exchange market for technology", to mainly raise the capability of indigenous innovation and push the development of the whole industrial chain. It was not difficult to exchange for the advanced technologies of foreign countries with the huge market size of China, but it is difficult to digest and absorb the imported technologies, make breakthrough in and master key technologies, and the upgrading of copying imitation to explorative innovation. In other words, we did not simply exchange market for technology, instead, we must master key technologies, and must increase the capability of indigenous innovation and continual innovation, to foster our own brands. At the very beginning of import, CHSR established the basic policy of "importing advanced technologies, joint design and manufacturing and making the brand of China". It not only helped China's High-Speed Railway industry build a modern manufacturing system, but also obtained the complete experience of product manufacturing and operation, to push the great development of the whole industrial chain of HSR at the minimum cost and in the shortest period of time.

Inspiration IV: give full play to the institutional advantages and main force role of state-owned enterprises in important fields and key sectors. In important fields related to the life line of national economy and key industries with large investment scale, long construction period and slow in obtaining returns, it is particularly important to give full play to the institutional advantages and main force role of state-owned enterprises. It is by relying on the institutional advantages that the large state-owned enterprises of the railway industry can, with the national strategy and national interests as the orientation, the major projects of the state as backup and the powerful organization and calling power as the guarantee, integrate the upstream and downstream of the huge industry chain, and realize the orderly cooperation and win-win benefit under the unified goal, to finally break through the major and key technical difficulties restricting the development of the industry, and establish Chinese standards through indigenous innovation, realizing the historical leap-forward from catch-up to taking the lead.

A.2 Speed up Making Breakthrough in Cutting Edge Technologies and Strategic Subversive Technologies in Rail Transit[2]

Today, rail transport has become a new growth driver to push forward new type urbanization in China and a new name card to go global.

On June 29, 2016, the executive meeting of the State Council adopted in principle the *Medium and Long-term Railway Network Planning,* which planned High-Speed railway network with the eight longitudinal lines including the coastal and Beijing-Shanghai lines and the eight transversal lines including the Continental Bridge Line and the line along the Yangtze River as the main, and supplemented by inter-city rails. In the meantime, with the in-depth implementation of the Belt and Road and HSR going global strategies, China is accelerating the construction of three transnational High-Speed railways of Eurasia High-Speed Railway, Central Asia High-Speed Railway and Trans-Asian High-Speed Railway. No doubt, the development of rail transport technologies in China has become a critical support to the promotion of new type urbanization and speeding up the going global of HSR.

At present, in infrastructure construction, rail transport is covering more quickly areas with more severe natural environment globally. Complicated marine environment and perilous mountain areas have been the important and difficult points in rail transport infrastructure construction, a variety of special geological and geomorphic units, ecologically fragile, extreme weather environment regions are even "prohibited areas" of its construction, in the related basic theory research, there are still many blank areas, there are obvious deficiencies with key technologies in scientific nature, economy and uncertainty, and problems of insufficient support to the decisions and actions in the HSR going global strategy are becoming daily prominent. In main equipment, independent design, domestic manufacturing and safe operation has been realized for the High-Speed MTU, communication signals and traction power supply equipment through import, digestion, absorbing and re-innovation, and realized the HSR products to Chinese standards. However, facing the challenges from continual innovation by powerful rail transport

[2]This is the introductory remarks written by the author of this book for the Journal of the China Railway Society No. 10, 2016.

countries, it is imperative for China to develop a new generation of HSR equipment and technologies that can run at higher speed, and are more economic, more environmental-friendly and safer, so as to maintain the competition advantages of CHSR in sustainable development with more competitive new technologies and more intelligent new products.

For example, High-Speed trains will have intelligent and adaptive bogie technologies, including the gauge-change technology required for HSR to go global, the light-weight car body technology with new materials, the high power electronic transformer technology, the permanent magnet motor and its traction control technology, full-electric braking technology, cophase power supply technology, energy saving transformer technology, equipment state monitoring technology, the application of new materials and new technologies, large capacity wireless communications and more accurate mobile blocking technology, intelligent transport organization and dynamic dispatching technology, and all these will be the core technologies for the development of the next generation of HSR. Furthermore, heavy loading technology with higher transport capacity, the renovation and upgrading technologies for existing railways will also be the emphasis of development of rail transport technologies.

On the other hand, rail transport has become diversified from the traditional trunk railway lines, urban mass transit has developed rapidly, and new technologies keep on emerging. In China, 100% low deck new type urban tram technology is in extensive application, and researches on the new technologies of energy creation, transfer and storage for power supply systems and new technologies of fuel cell, super capacitor and wireless energy transfer are quite active. Straddle or hanging type monorail train technology that can avoid interference with ground traffic, the automatic running "capsules" train technology for the "last kilometer" of mass transit, and the low cost virtual railway train technology has attracted wide public attention. With the operation of the maglev train on Changsha Airport Line, people place infinite expectation on maglev train, and the state has included in the advanced mass transit for the 13th Five-Year Plan the research on the medium speed maglev transport system with speed of 200 km/h and on high speed maglev transport system with maximum design speed of 600 km/h.

The manned test of low temperature superconducting maglev train at a speed of 603 km/h and the start of construction of the Tokyo to Nagoya (finally to Osaka) operation line in Japan has proved the possibility to build mass transit with higher speed using the superconducting maglev technology, and the high temperature superconducting maglev train technology with proprietary intellectual property right of China, with its unique self-suspension, self-steering and self-stability performance, has become the best pattern for the ground super-high speed rail transport at a speed of about 1000 km/h. Furthermore, the ground super-high speed mass transit in vacuum pipe is paid much attention, Southwest Jiaotong University created the first vacuum pipe high temperature superconducting maglev transport experimental system in the world in 2014, attracting extensive attention in the international academic circle, and it was called "super chute" by the flagship journal Spectrum of IEEE. Recently, as the United States proposed the concept of

Hyperloop based on vacuum pipe and its engineering research, the ground super-high speed mass transit approaching the sonic speed is coming closer to us.

Mass transit technologies are upgrading by generations and changing with each passing day. We have the reason to believe that, the *Journal of the China Railway Society*, as the highest academic exchange platform in the mass transit field of China, will surely become the "sowing machine" and "publicity team" for the next generation of HSR technologies, new type urban mass transit technologies, maglev transport delivery and vacuum pipe transport technologies.

A.3 A National Level Headquarters Is Required for CHSR to Go Global[3]

—HSR going global will start a new land right era with right of way supporting the land right.

From June 16 to 19, 2014, Premier Li Keqiang of the State Council paid an official visit to the UK, and held the annual meeting of the prime ministers of the two countries. Remarkably, the topic of High-Speed rail cooperation was included in a joint statement issued by the two countries. It was also the fourth time that Li Keqiang recommended China's High-Speed rail to foreign countries in his visits.

More than that, recently, the construction of phase II of the main body works of the High-Speed railway between Ankara and Istanbul in Turkey has been completed, and the Ethiopia-Djibouti Railway, constructed completely to the Chinese standards, has started rail paving officially. Messages reported by media indicate that the pace for CHSR going global is being accelerated.

Recently, *China Youth Daily* interviewed professor Fei Xu, President of Southwest Jiaotong University. He believed that CHSR was going global in a very rare important period of strategic opportunities, it would bring China a new era with land right supported by the right of way, and help China march on a road of national security with Chinese characteristics.

China Youth Daily: What is the special significance of the frequent presence of CHSR on the stages of opening to the outside world and diplomacy?

Fei Xu: In recent years, we have been paying increasing attention to the "marine right", but in the new security situation that the opening to the east (the Pacific) has encountered the containment by the United States, it is necessary to re-examine the relationship of the "marine right" and the "land right", to establish a new land right to hedge the marine right.

[3]This article was published on China Youth News dated July 10, 2014 on p. 7, as a special interview of the journalist of the newspaper to the author.

China should, while firmly implementing the deep blue strategy, take the HSR as the mainstay to break the siege to gain access to the Atlantic Ocean in the west and to the Indian Ocean in the south. This will establish the new land right era with the right of way supporting the land right, to ensure the overall security of the state, and help China get a road of national security with Chinese characteristics. In the meantime, based on the existing and daily enhancing marine right in the Pacific Ocean, we can build up the new marine right based on the Indian Ocean, the Atlantic Ocean and even the Arctic Ocean.

Therefore, the HSR going global maintains China's strategic position in the geopolitics of Eurasia and the global landscape, and carries the major missions of international security, economic security, political security and national security.

China Youth Daily: How to realize supporting the "land right" with the "right of way"?

Fei Xu: In the long-term space layout, it should cover three dimensions. One is continuing the planning and consultation, and speeding up the construction of four strategic lines, i.e. the "Eurasian High-Speed Railway", starting from London, running through Paris, Berlin, Warsaw, Kiev and Moscow to Urumqi in China, and finally ending at Beijing. The Central Asia High-Speed Railway, which will start from Kashgar of Xinjiang, cross Kazakhstan, Uzbekistan, Turkmenistan, Iran, Turkey and other countries, enter Europe via Bulgaria and finally end in Germany. The "Trans-Asian High-Speed Railway", starting from Kunming, and running via Cambodia, Thailand and Malaysia to end in Singapore. The China-Pakistan Railway, starting from Kashgar of Xinjiang via Islamabad to Karachi and Gwadar Port, to reach the Indian Ocean.

The second is to accelerate the construction of railways in Central Africa. Two lines can be planned: from Saudi Arabia, Yemen and Dubai to Ethiopia in Africa; or from Kuwait, Iraq, Syria (Damascus), Jordan via Suez Canal of Egypt to Africa.

The third is accelerating the study of China-Russia-Canada-United States High-Speed Railway. It will start from the northeast of China, run through the Siberia to the Bering Strait, cross the Pacific Ocean with a tunnel across the ocean to Alaska, then from Alaska to Canada, and finally to the United States.

China Youth Daily: Germany, France, Japan and Canada are traditional powerful countries in High-Speed railway technologies, remaining strongly competitive on the international market. As a late-comer, what advantages the CHSR has on the international market?

Fei Xu: As far as foreign environment is concerned, the global economy has further recovered, the construction of HSR in North America has started, the Eurasia Continent economic corridor now plays a prominent role, and there is booming demand for the modernization of railways in Africa. As far as our own conditions are concerned, China has accelerated the upgrading of industrial structure, and enhanced the complementarity with the development of relevant industries of the state; the transportation equipment of China features complete and sound technologies, mature operation experience and high quality and performance-to-price ratio, enjoying good reputation on the international market.

China Youth Daily: Then, where are the challenges?
Fei Xu: CHSR going to the international market faces unusual difficulties and serious challenges in financing, technical standards and transnational operation, and is also subjected to the complicated effects of environmental factors such as international politics, market, industries, laws, finance and culture.

First, it faces fierce competition from developed countries, such as Germany, France, Japan and Canada, which already have the advantages of High-Speed railway technology. The great concerns of relevant countries about their own factors such as national security, resources protection and enterprise interests also have caused many difficulties in the negotiations for cooperation. For example, the website of "The Hindu" published an article, saying "India should say no to China High-Speed railway".

Technical standards have become a technical barrier in international trade. In the background that HSR standards have already been monopolized or used by foreign countries, and European standards have the natural exclusiveness, the core technologies of CHSR can be smoothly applied worldwide only after obtaining international patent and property right protection.

In addition, the investment and financing and operation patterns are not quite mature, the BT, BOT, EPC and PPP patterns have their respective advantages and disadvantages, with risks in profiting. In some developing countries, the political situation is not stable, with religion conflicts and repelling cultures, these are also uncontrollable factors in CHSR going global.

China Youth Daily: How should we cope with these challenges?
Fei Xu: It is suggested that the state should work out countermeasures in five aspects: commanding the implementation of strategy, gathering forces of the industry, building a fostering and training system, establishing a national think tank, and innovating investment and financing methods.

HSR projects are big deals, and negotiations cannot be held only between enterprises. It must be governmental action, therefore there should be a headquarters at the national level—the national strategic committee for HSR going global. Government takes the lead, to push it forward by forming consortiums with division of work, to coordinate all parties with planning in advance in various dimensions of politics, economy, industries, equipment, investment and financing, diplomacy, national defense and culture, and to provide time table and roadmap for the implementation of the national strategy of HSR going global. Also, this should be used as platform to further strengthen dialogue and consultation between governments, and establish normal communication and coordination mechanism with target countries.

To reduce the excessive and bad competition between domestic enterprises, eliminate internal consumption to go to international market with concerted efforts, and participate in overseas competition in groups, the international industrial alliance for CHSR going global should be established as early as possible, to form an organizational collaboration system covering the full life cycle of HSR industry. Especially, the establishment and completion of internationalized HSR standard system of China should be accelerated, to push forward the development and recommendation of HSR standards of China.

CHSR going global is only our will, and it is absolute principle to enter target countries. For the CHSR to go global, we should also first train and foster talents, and build up the talents "pool" for continual going global.

A high level national think tank is also required for CHSR to go global. It is suggested to build worldwide development and research bases for HSR to go global, to provide all-round "think tank" service for the going global strategy of CHSR, and establish regional development and research centers for HSR to go global. Africa, as an important pole on the world political stage, a new pole of global economic growth and a colorful pole of human civilization, will be important outposts for CHSR to go global. We can establish the "Africa HSR R&D Center" at appropriate time, to provide powerful perspective intelligence support for CHSR to land on African continent.

We should also explore stable and efficient ways of investing and financing to cope with the risks and challenges posed by unfamiliar environment for HSR going global. The specific ways include strengthening the bank-enterprise cooperation, flexible application of investment and financing patterns, improving financing ability on the international financial market, research and spreading of the "resources for High-Speed rail" pattern, and accelerating the construction of international banks.

In his visit to the UK, Premier Li Keqiang talked about the HSR project from the north of UK to London. It would be of epoch-making milestone significance if China could build HSR in the UK, once an empire that the sun never set, because the world's first railway was born in Britain. This will also announce to the whole world that China not only produces shoes, stocks and lighters at the low end of the industrial chain, but also builds High-Speed railways.

A.4 Re-practice of Mao Yisheng's Engineering Education Concept[4]

The year 2016 was the 120 anniversary of the birth of Mr. Mao Yisheng, a great scientist of China.

Mr. Mao devoted all his life to the bridge and railway undertakings of China. Before the War of Resistance against Japanese Aggression, he took charge in the first modern bridge designed and constructed by the Chinese – Qiantang River Bridge. Only 89 days after the bridge was open for traffic, to resist against the

[4]This is the signed article published by the author of this book on Guangming Daily dated March 1, 2016 on p. 14.

invasion by the Japanese imperialists, Mr. Mao arranged in person to blast off the bridge built by him, and wrote a poem in the night of blasting the bridge, to express the patriotic determination to restore the bridge. After the victory of the war, Mr. Mao had the Qiantang River Bridge re-built in a very short period of time. The bridge was constructed, blasted and constructed again in 14 years, condensing the patriotic sentiments of Mr. Mao to devote himself to the nation.

Mr. Mao was admitted to the Civil Works Department of Tangshan Industrial Special School (predecessor of the Southwest Jiaotong University) of the Ministry of Transport in 1911, with the student No. 393. He graduated in 1916 with the total average score of 92, ranking the first in the grade. He went to study in Cornell University in 1917 as No. 1 in the examination for postgraduates officially financed by the US government, became the first doctor of engineering of the Engineering College of the Carnegie Mellon Institute of Technology in 1919, and resolutely came back to his alma mater to teach. Mr. Mao was back to the alma mater in 1920, became a professor at the age of 24, and was appointed the 21st president of the school before he was 30 years old, as one of the youngest university professors and presidents at that time. Mr. Mao said when recalling his school life: "the efforts of these 14 years were like erecting a bridge, building solid piers for the cause of my whole life."

Mr. Mao was appointed president of the alma mater four times, especially in the four years when he led the school during the War of Resistance against Japanese Aggression, he did his utmost to keep the school running, and arranged over ten relocations over journeys of several thousand kilometers, finally returning to Tangshan. We are unable to express in words our gratitude to Mr. Mao for his enthusiasm and devotion to his alma mater, and his teaching ideas are even more our precious wealth.

"Learn before practice"

Mr. Mao was engaged in teaching and education reform all his life, especially, he paid great attention to higher engineering education, and the most commended is his education concept "learn before practice". In his own words, the idea "learn before practice" came from "study the nature of things in order to acquire knowledge" in the *Great Learning*. This just corresponded to the modern constructive learning theory.

This idea "learn before practice" is shining with the light of wisdom today when higher education is undergoing fundamental changes. The dispute on whether knowledge comes from outside or inside has remained since the ancient times about the source of knowledge and the methods of acquiring knowledge. The constructivism that integrates the essence of both believes that learning is essentially process of individuals constructing their new knowledge and understanding on the basis of the existing experience and knowledge through interaction with the outside world. Therefore, students are the subjects of learning, and are active knowledge discoverers and constructors.

Early in the first half of the 20th century, Mr. Mao noticed with acute eyesight problems existing in higher education. He pointed out that the education "put students at a passive position, only to give them knowledge in a spoon-feeding manner, with the danger of doctrinarism and reaching for what is beyond grasp". Therefore he opposed the education method of "only teaching knowledge at classes. Ignoring the aspects of how to enlighten the creativity of students, to develop their intelligence, and to identify their personality. As a result, students only become mechanical as they receive more knowledge". Obviously, Mr. Mao Yisheng did profound thinking on "teaching" and "learning", demonstrating the modern education concept of "centering on students and study". It is the teaching practices to promote this education concept that Southwest Jiaotong University today emphasizes in its "four-in-one" innovative talents fostering system "exploring knowledge" instead of "knowledge transference".

A wider base can support a higher pyramid in building a pyramid with stones

As an engineering educationist, Mr. Mao did not neglect the overall education and fostering of students because of focusing on engineering education. He pointed out in an article in 1981: "those learning literal arts should learn some sciences, and those learning sciences should learn some literal arts. All people can learn some music and fine arts." He also advised young people with the example that Marx and Engels had an ardent love for mathematics: "One should first have a wide rank of knowledge before becoming an expert, and those want to become experts in a specific discipline should never restrict their vision only within the scope of that discipline." In his view, there is no fundamental conflict between "general" and "specializing", and they are in the relationship like "a wider base can support a higher pyramid in building a pyramid with stones".

In this era of innovation, Southwest Jiaotong University hopes to explore on wide scope specialized education based on exploring ways of general education, instead of simply inheriting the specialized education of the Soviet Union or the general education of the United States, to well implement the education idea of Mr. Mao Yisheng in the teaching management, and deeply integrate general education with specialized education.

Based on the opinions of most engineers, Mr. Mao summed up the successful factors of engineers into six aspects, and arranged them by their importance as: conduct, determination, agility, social accomplishment, knowledge and skills. In our opinion, these factors should be fostered in five classes, i.e. the traditional and current classroom education and the derived after-class education, as the first class; the education experience of off-class activities on culture, science and technology, arts and sports, social groups, public welfare and interesting groups, as the second class; education with practical activities such as social practices, internship, practical training and volunteering work, as the third class; internationalized interactive learning, including building up the international education concept, construction and use of international teaching resources, international cooperative school running, and creating and experiencing internationalized education environment, as the fourth class; and virtual classroom and network interactive learning such as E-learning, or MOOC courses, E-class, cloud study and flipped classroom, as the fifth class.

The "five classes" will push forward talents fostering in collaboration, with the main goal to build the innovative talents fostering system with the characteristics of

SWJTU that integrating "value moulding, personality cultivation, ability fostering and knowledge exploration, the core element is the talents fostering plan integrating the five classes, and basic method is to establish the fostering system with the "five classes" collaborating and integrating, and embodying the full-process of fostering, to realize the overall and in-depth integration of "five classes".

Mr. Mao said: "a lot of knowledge are inter-connected. To acquire deep knowledge and achieve something in a certain aspect, you must first learn widely, to get foundation in many aspects……some knowledge may seem irrelevant to your own specialty, but if you have learned it to widen your eyesight, it can inspire your thinking, deepen your understanding of knowledge and promote study." He also stated that "education for students should be based on the principle of wide rank of knowledge and sufficient ability to learn", we should "arrange more disciplines related to humanity and culture", and overcome the defect that "lessons for students are restricted by physical things, lacking in culture education that can cleanse the spirit, resulting in narrow-mindedness, and illiteracy to personnel affairs, economy, action and machinery". This is typical trans-discipline education idea.

To strengthen trans-discipline fostering, we started pilot in the Mao Yisheng College, to build up trans-discipline education platform with trans-discipline specialties, trans-discipline curriculums, trans-discipline teaching teams and trans-discipline classes. In 2015, on the basis of the research-oriented top-notch innovation class, we established trans-discipline composite elite classes covering areas of engineering, sciences, literature and finance, and adopted the form of combining broad enrollment with broad fostering and combining trans-discipline study and personality fostering, so that students can study trans-discipline in a number of discipline combinations by selecting a major and a minor program according to their interests and strong points. In 2016, the school established the first batch of trans-discipline curriculum, with teaching team covering disciplines of engineering, sciences, literature and finance, and a number of areas of teaching, scientific research, management and enterprises; the classes are also attended by students of various specialties of the whole school.

In his period of teaching, Mr. Mao often invited students home for talks after class, to know the students' mastery and understanding of classes in their unfettered talks. The first academy set up in Southwest Jiaotong University was name after the old president Mao Yisheng—Tang Chen Academy. The academy is a courtyard with books in form, but its connotation is far more than the form as seen. It is immersed in the literacy fragrance, filled with wisdom and permeated with passion, and power is growing here. It is important to have good teachers in learning, and the essence of education should exist between teachers and students.

We hope that the people of SWJTU will pass from generation to generation the kind virtue, intellectual cultivation, the passion for the family and country, the style of dedication to work, the character of seeking truth and the courage of innovative opening up of Mr. Mao Yisheng. With the academy as a carrier, we will promote smooth exchanges between teachers and students and between students, so that the pursuit for knowledge and cultivation of virtue will go with the students in every step of their growth.

Postscript

This book is based on my speeches at a number of forums and conferences. Since the second half of 2013, I attended conferences and forums such as the Summit Forum on CHSR Going Global, Strategic Symposium on CHSR Going Global and Development, Forum on Belt and Road Rail Transport Development and Cooperation, Strategic Forum on Silk Road Economic Belt and CHSR Going Global, TOD & PPP International Symposium on Rail Transport, the 9th World HSR Conference (Tokyo), and the 2016 Global University Presidents Summit Forum, Summit Forum on China-Europe Higher Education and Humanistic Exchange, China-Russia University Presidents Summit Forum (Moscow), the Fourth China-India Strategic and Economic Dialogue (New Delhi), the 25th Plenum of Affiliated Higher School Working Consultancy Committee of the Ministry of Education and China Higher Engineering Education Summit, and invited to deliver keynote and them speeches. The topics covered the Belt and Road, HSR going global, the global strategic value of CHSR, the global strategy of CHSR, the future of HSR and engineering education in universities.

It it the promotion by Ms. Xin Yanxiang of the Truth and Wisdom Press that has made it possible to present this book to readers as it is now. The publication of this book will undoubtedly be much later without her efforts and assistance. I know from her letters to me that she got the idea to publish a book early in 2013 when she noted some of my articles about HSR in WeChat. In fact, we have maintained contact since 2010 when Ms. Xin was the responsible editor for the book *Generous without a Corner* for which I was the editor-in-chief, and later I published *Power of Culture* and *Demonstration of the Stability, Defectives and Evolution of Strategic Alliances* at the Truth and Wisdom Press. At the end of 2016, she assisted me to re-arrange nearly 20 speeches about Belt and Road and HSR strategy and articles published on *Frontiers of Academia, Engineering Sciences* and *Guangming Daily* into chapters as a complete system, as the initial form of this book. On this basis, I finished the book mainly in the winter vacation this year by making supplements,

F. Xu, *The Belt and Road*, https://doi.org/10.1007/978-981-13-1105-5

combination and refining, optimizing the structure and polishing it. In this whole process, she provided many constructive modification suggestions, and more than that, she checked the whole book and helped me draw all figures and tables. Furthermore, she deliberated repeatedly on the title of the book, and did careful and refined design of the cover and the size and font for the book to make it perfect. I am quite grateful to Ms. Xin for her excellent editing work, and her intelligent contribution and hard work!

In more than three years when I worked in Southwest Jiaotong University, I got the care, assistance and support from many colleagues in the school. Exchange of views with them is an extremely important chance of learning to me. Many of them are the most outstanding scholars and scientific and technological workers in the rail transport field of China and the world, learning from, and exchange and consultation with them have benefited me a lot. I would specially thank academicians Shen Zhiyun, Qian Qingquan and Zhai Wanming of Southwest Jiaotong University, and thank professors Zhang Weihua, He Chuan and Luo Shihui, and Professor Gao Bai, director of China HSR Development Strategic Research Center of SWJTU and professor Xiang Hong, dean of Belt and Road Development Research Institute. There are still many more colleagues to whom I want to express my thanks, just forgive me that I cannot list their names one by one. The best return to them is to speed up the pace to build a first-rate university with distinctive transportation features, and realize on an early date the dream of SWJTU "a team of many masters, talents keeping on emerging, outstanding contribution and ever-lasting cause".

In recent years, I have participated in many BBS activities related to the Belt and Road, and I have benefited a lot from the discussions with many scholars and officials at home and abroad in various fields. I want to express my special thanks to Professor Li Wuwei, Vice Chairman of the 11th National Committee of CPPCC, to Mr. Yao Peisheng, former Deputy General Director of the Department of European-Asian Affairs and Counsellor Gao Zhenting of the Department of International Economy of the Ministry of Foreign Affairs, to General Director Wang Ming of the Comprehensive Transportation Research Institute and General Director Ao Wangzhong of the Regional Cooperation and Development Research Institute of International Cooperation Center of National Development and Reform Commission, and Division Head He Jiangchuan of the Administrative Office of State Leading Group for Promoting the Belt and Road Construction Work, to Director Liu Huaqin of the Europe Research Department of the International Trade and Economic Cooperation Research Institute and Dr. Wu Yabin of the International Trade Negotiation Representative Office of the Ministry of Commerce, to Major General Xiao Yusheng of the Academy of Military Sciences of the PLA, Major General Qiao Liang of the Strategy Teaching and Research Office of Air Force Command College, to Director Xue Li of the International Strategy Office of World Economics and Politics Research Institute of Chinese Academy of Social Sciences and Secretary Li Houqiang of the Academy of Social Sciences of Sichuan Province.

I also extend my many thanks to deputy editors-in-chief Shen Weixing and Liu Wei of the *Guangming Daily*. My sincere thanks to Sir Alan, President of Leeds University, UK, Professor Peng Long, President of Beijing Foreign Studies University, Professor Zhu Xiaoming, former Dean of China Europe International Business School, and Professor, M. Rapik Saat of the Railway Research Center of the University of Illinois at Urbana-Champaign (UIUC), USA. I also want to thank Deputy Secretary General Zong Wa of China Education Association for International Exchange and President Feng Yaowu of China-EU Association. Thanks to Director Patrick Duong of the Regional Program of Asia-Pacific Regional Center Governance Department and Gu Qing, Head of the Poverty Alleviation, Equity and Governance Division of the UNDP, and Ms. Izabela Kalinowska, chief delegate at the Chengdu Office of Lodz Province and Lodz City Government of Poland. In particular, I would like to thank Consul General Hasan Habib of the General Consulate of Pakistan to Chengdu and Consul General Deniz Eke of the General Consulate of Turkey to Shanghai.

In these years, I attended a number of symposiums and forums and participated in subject researches on CHSR going global, and had the opportunities to exchange views with the leaders, experts and peers of the State Railway Administration, China Railways Corporation, railway schools and scientific research institutes, main railway enterprises and the UIC, and many views in this book are based on inspirations from discussions with them. Therefore I would like to express my sincere thanks to Academician Fu Zhihuan, former Minister of Railways, Academician Sun Yongfu, Vice Minister of Railways and Lu Chunfang, President of the Council of China Railway Society, to Dr. Yang Yudong, Deputy Minister of Ministry of Transport and General Director of State Railway Administration, Mr. Zheng Jian, member of the Leading Party Group and Yan Hexiang, General Director of Science and Technology and Law Department of State Railway Administration, and to Lu Dongfu, General Manager of and Chief Engineer and Academician He Huawu of China Railways Corporation. I express my deep thanks to Academician Tian Hongqi, Deputy President of Chinese Academy of Engineering, Academician Wang Mengxu of Beijing Jiaotong University and Academician Du Yanliang of Shijiazhuang Railway University, to President Xi Guohua, Vice President Wang Jun, Chief Engineer Zhang Xinning and Chief Information Engineer Wang Yongzhi of CRRC, and to Chairman Li Changjin of China Railways Corporation and General Manager Huang Difu of China Railways International Co., Ltd. I extend my sincere thanks to Academicians Liu Youmei and Ding Rongjun of CRRC Zhuzhou Electric Locomotive Company, and Academician Qin Shunquan of China Railway Bridge Bureau, to Dong Shouqing, Deputy General Director of China Academy of Railway Sciences, Zhu Ying, General Manager of China Railway Eryuan Engineering Group, Wang Yuze, Deputy General Director of China Railway Siyuan Survey and Design Group and Lin Zhonghong, Deputy General Director of China Railway Economic and Planning Research Institute. Finally, I would like to thank Mr. Jean-Pierre Loubinoux, Secretary General of UIC and Mr. Ignacio Barron, Director of Passenger Transport and High-Speed Railway Technical Department of UIC.

Of course, I will be responsible for any error or improper expression in this book, and they have nothing to do with those listed above.

Needless to say, the understanding, encouragement and love of my family are the biggest motive force and source of happiness in my work and life. My parents and other people of my family are quite kind, and tolerate my negligence of care and greetings to them due to my busy work, and my bright daughter keeping on progress also makes me quite pleased. My wife does almost all housework, and gives me meticulous care and love, so that I can concentrate in work and writing without distraction, which is of great importance to the completion of this book.

When this books was about to be sent to the printer, the first Belt and Road Forum for International Cooperation was held in Beijing from May 14 to 15, 2017. 29 foreign heads of state and heads of government, as well as 3 responsible persons of important international organizations including the Secretary General of the United Nations, Chairman of the International Committee of the Red Cross, attend the forum, and about 1500 distinguished guests from all walks of life from more than 130 countries were present at the forum as official representatives, with unprecedented magnitude and influence. The Chinese Government also declared to hold the second Belt and Road Forum for International Cooperation in 2019. For more than three years since President Xi Jinping put forward the Belt and Road Initiative in autumn 2013, this initiative has not only bestowed new connotation of the times to the ancient Silk Road, but also become a historical creation and "century program" in the opening to the outside world of China in the new situation.

For the development of the Belt and Road Initiative in the new historical horizon, we should join hands with all countries and areas, make use of potential advantages, and speed up the construction of network with infrastructures including High-Speed railways, to realize connectivity and cooperation, and make greater contribution to the building of the community with a shared future for mankind.

Fei Xu
June 2017

MIX
Papier aus verantwortungsvollen Quellen
Paper from responsible sources
FSC® C105338

If you have any concerns about our products,
you can contact us on
ProductSafety@springernature.com

In case Publisher is established outside the EU,
the EU authorized representative is:
Springer Nature Customer Service Center GmbH
Europaplatz 3, 69115 Heidelberg, Germany

Printed by Libri Plureos GmbH
in Hamburg, Germany